Algebraic Number Theory

SECOND EDITION

CHAPMAN AND HALL
MATHEMATICS SERIES

Edited by Professor R. Brown
Head of the Department of Pure Mathematics
University College of North Wales, Bangor

A First Course on Complex Functions
G. J. O. Jameson

Rings, Modules and Linear Algebra
B. Hartley and T. O. Hawkes

Galois Theory
Ian Stewart

Introduction to Optimization Methods
P. R. Adby and M. A. H. Dempster

Graphs, Surfaces and Homology
P. J. Giblin

Linear Estimation and Stochastic Control
M. H. A. Davis

Mathematical Programming and Control Theory
B. D. Craven

Ordinary Differential Equations
D. K. Arrowsmith and C. M. Place

Independence Theory in Combinatorics
V. Bryant and H. Perfect

Introduction to Algebraic K-Theory
J. R. Silvester

Algebraic Number Theory

SECOND EDITION

I.N. STEWART
D.O. TALL
Mathematics Institute
University of Warwick
Coventry

CHAPMAN & HALL/CRC

Boca Raton London New York Washington, D.C.

Library of Congress Cataloging-in-Publication Data

Catalog record is available from the Library of Congress.

First edition 1979
Second edition 1987
Reprinted 1987, 1992
First CRC Press reprint 2000
Originally published by Chapman & Hall

International Standard Book Number 0-412-29690-X (PB) 1 2 3 4 5 6 7 8 9 0
International Standard Book Number 0-412-29870-8 (HB) 1 2 3 4 5 6 7 8 9 0
Printed in the United States of America
Printed on acid-free paper

TO RONNIE BROWN
WHOSE BRAINCHILD
IT WAS

Contents

PART III NUMBER-THEORETIC APPLICATIONS

Preface

The title of this book may be read in two ways. One is 'algebraic number-theory', that is, the theory of numbers viewed algebraically; the other, 'algebraic-number theory', the study of algebraic numbers. Both readings are compatible with our aims, and both are perhaps misleading. Misleading, because a proper coverage of either topic would require more space than is available, and demand more of the reader than we wish to; compatible, because our aim is to illustrate how some of the basic notions of the theory of algebraic numbers may be applied to problems in number theory.

Algebra is an easy subject to compartmentalize, with topics such as 'groups', 'rings' or 'modules' being taught in comparative isolation. Many students view it this way. While it would be easy to exaggerate this tendency, it is not an especially desirable one. The leading mathematicians of the nineteenth and early twentieth centuries developed and used most of the basic results and techniques of linear algebra for perhaps a hundred years, without ever defining an abstract vector space: nor is there anything to suggest that they suffered thereby. This historical fact may indicate that abstraction is not always as necessary as one commonly imagines; on the other hand the axiomatization of mathematics has led to enormous organizational and conceptual gains.

Algebraic number theory illustrates both of these tendencies, and the 'creative tension' engendered by their overt opposition (and covert collaboration); and we hope that a study of it will encourage an awareness in the student that it is possible to use abstract algebra to prove theorems about something *else*. A particular target here is a partial proof of Fermat's Last Theorem, chosen for its historical importance and cultural notoriety, which add a certain piquancy to its excellence as a motive for the introduction of a number of key ideas. Fermat stated that the equation

$$x^n + y^n = z^n$$

has no non-zero integer solutions for x, y, and z, when n is an integer greater than or equal to 3. We shall prove this, following the original ideas of Kummer, for the case where n is a 'regular' prime; but under the simplifying assumption that n does not divide any of x, y, or z.

Our algebra will not be as abstract as it might be. Professionals may be aghast to learn that Galois group theory is not used, that valuations make no appearance, and that generalizations such as Dedekind domains have been suppressed. Instead, we make use of such 'classical' (a less polite word would be 'oldfashioned') devices as symmetric polynomials and determinants. The reason is that the latter are accessible to a much wider readership. For the same reason we have devoted more space to some aspects of the subject-matter (notably factorization into primes) than current importance would warrant: these topics provide ample opportunity for concrete computations which help familiarize the student with the underlying concepts. (In a lecture course some of this material could of course be omitted.) In mathematics it is important to 'get one's hands dirty', and the elegance of polished theories does not always provide suitable material. Prime factorization in specific number fields also displays the tendency of mathematical objects to take on a life of their own: sometimes something works, sometimes it does not, and the reasons why are far from obvious. Of such frustrating yet stimulating stuff is the mathematical fabric woven.

Of necessity we must assume *some* algebraic background. The reader will be expected to have a working knowledge of rings and fields, ideals and quotient rings, factorization of polynomials, field extensions, symmetric polynomials, the idea of a module, and free abelian groups. This material is revised in detail in Chapter 1. Apart from this we shall assume only some elementary results from the theory of numbers and a superficial comprehension of multiple integrals.

For organizational convenience, rather than mathematical necessity, the book is divided into three parts. The first develops the basic theory from an algebraic standpoint, introducing the ring of integers of a number field and exploring factorization within it. Quadratic and cyclotomic fields are investigated in more detail, and the Euclidean imaginary quadratic fields are classified. We note that factorization into irreducibles is not always unique in a number field, but that useful sufficient conditions for uniqueness may be found. The factorization theory of *ideals* in a ring of algebraic integers is more satisfactory, in that every ideal is a unique product of prime ideals. The extent to which factorization of elements is not unique can be 'measured' by the group of ideal classes (fractional ideals modulo principal ones).

The second part emphasizes the power of geometric methods arising from Minkowski's theorem on convex sets relative to a lattice. We prove this geometrically by looking at the torus that appears as a quotient of Euclidean space by the lattice concerned. As illustrations of these ideas we prove the two- and four-squares theorems; as the main application we prove the finiteness of the class group.

The third part concentrates on applications of the theory thus far developed, beginning with some slightly *ad hoc* computational techniques for class numbers, and leading up to the aforementioned special case of Fermat's Last Theorem. We also prove Dirichlet's theorem on the finite generation of the group of units, which strictly speaking is not an application but a part of the basic theory–but it fits best at this point.

A final appendix deals with quadratic residues and the Quadratic Reciprocity Theorem of Gauss. It uses straightforward computational techniques (deceptively so: the ideas are very clever) and is placed separately because it has a different flavour from the rest of the text. It may be read at an early stage – for example, right at the beginning – or alongside Chapter 3, which is rather short: the two together would provide a block of work comparable with the remaining chapters in the first part of the book. A certain amount of historical material is included for motivation and light relief.

A preliminary version of this book was written in 1974 by Ian Stewart at the University of Tübingen, under the auspices of the Alexander von Humboldt Foundation. This was used as the basis of a course to students at the University of Warwick in 1975, and subsequently expanded (despite strenuous efforts to keep it short) into the present volume. It has been much improved by the subtle comments of a perceptive, but anonymous referee; by the admirable persistence of students attending the course; and by discussion with colleagues in the Mathematics Institute at the University of Warwick. For this second edition we have brought the historical material up to date, added some detailed examples, removed superfluous sections and corrected misprints. In particular we are grateful to M.H. Eggar and Akira Takaku, who sent us lists of typographical and computational errors; and to several generations of students at Warwick who have used the first edition of the book for a reading course, and made many constructive criticisms. No blame may be attached to them for any errors: the infelicities that remain may very well be the responsibility of our co-author.

Coventry, January 1986.

Ian Stewart
David Tall

Readers' guide

Theorems, lemmas, propositions, and corollaries are numbered consecutively in the form **m.n** where **m** is the number of the chapter. Displayed equations are numbered consecutively within each chapter in the form (n). References [n] are to the books and papers listed on page 250. Two commonly occurring references for background material are identified as [HH] and [GT]. The end (or absence) of a proof is signalled by the symbol '□'.

Structure of the book

The logical dependence of chapters is approximately as shown in the diagram on the next page.

For a course providing the minimum material needed to prove Theorem 11.8 (a special case of Fermat's conjecture) the following sections may be omitted: 2.6, 3.1, 4.7–4.9, 5.4, 7.2, 7.3, 9.4, 9.5, 10.2–10.4, 12.1–12.4. A course aimed at Dirichlet's units theorem could omit 3.2, 4.6–4.9, 7.2, 7.3, 9.3–9.5 and the whole of Chapters 10 and 11. In neither instance is the Appendix necessary.

In planning a regular schedule of work it should be noted that the chapters in Parts II and III are generally shorter than those of Part I, and two (or even more in the case of Chapters

6-8) may be taken together to give a block of work comparable with the earlier chapters. In particular, Chapters 6, 7 and 8, 9 may be taken as blocks in the schedule; and putting the Appendix on Quadratic Reciprocity in one week alongside Chapter 3, this breaks the text into ten parts for a ten week, one term course. For a slower-paced 15-week course, Chapters 1, 2, 4 and 11 could be divided into two, and Chapters 8 and 9 kept separate.

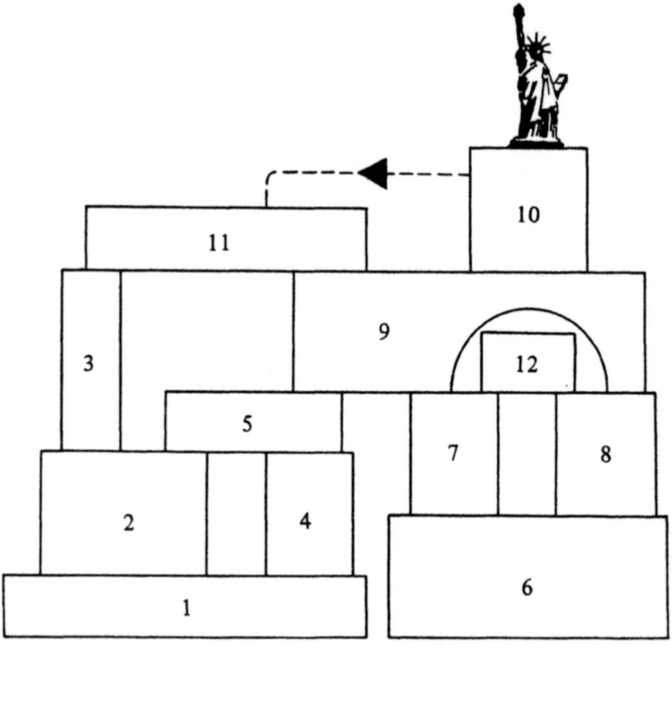

Index of notation

The origins of algebraic number theory

Numbers have fascinated civilized man for millenia. The Pythagoreans studied many properties of the natural numbers 1, 2, 3, . . . , and the famous theorem of Pythagoras, though geometrical, has a pronounced number-theoretic content. Earlier Babylonian civilizations had noted empirically many so-called Pythagorean triads: 3, 4, 5; 5, 12, 13. These are natural numbers a, b, c such that

$$a^2 + b^2 = c^2. \tag{1}$$

A clay tablet from about 1500 B.C. includes the triple 4961, 6480, 8161, demonstrating the sophisticated techniques of the Babylonians.

The Ancient Greeks, though concentrating on geometry, continued to take interest in numbers. In *c.* 250 A.D. Diophantus of Alexandria wrote a significant treatise on polynomial equations which studied solutions in fractions. Particular cases of these equations with natural number solutions have been called *Diophantine* equations to this day.

The study of algebra developed over the centuries too. The Hindu mathematicians dealt with increasing confidence with negative numbers and zero. Meanwhile the Moslems conquered Alexandria in the 7th century, sweeping across north Africa and Spain. The ensuing civilization brought an enrichment of mathematics with Moslem ingenuity grafted on to Greek and Hindu influence. The word 'algebra' itself derives from the arabic title 'al jabr w'al muqabalah' (literally 'restoration and equivalence') of a book written by Al-Khowarizmi in *c.* 825. Peaceful co-existence of Moslem and Christian led to the availability of most Greek and Arabic classics in Latin translations by the 13th century.

In the 16th century, Cardano used negative and imaginary solutions in his famous book *Ars Magna,* and in succeeding centuries complex numbers were used with greater understanding and flexibility.

Meanwhile the theory of natural numbers was not neglected. One of the greatest number theorists of the 17th century was Pierre de Fermat (1601–1665). His fame rests on his correspondence with other mathematicians, for he published very little. He would set challenges in number theory based on his own calculations; and at his death he left a number of theorems whose proofs were known, if at all, only to himself. The most notorious of these was a marginal note in his own personal copy of Diophantus, written in Latin, which translates:

To resolve a cube into [the sum of] two cubes, a fourth power into fourth powers, or in general any power higher than the second into two of the same kind, is impossible; of which fact I have found a remarkable proof. The margin is too small to contain it . . .

More precisely, Fermat asserted, in strong contradistinction to the case of Pythagorean triads, that the equation

$$x^n + y^n = z^n \tag{2}$$

has *no* integer solutions x, y, z (other than the trivial ones with $x, y,$ or z equal to zero) if n is an integer greater than or equal to 3. To date this assertion has been proved only for certain values of n – admittedly larger in number, and including all $n \leqslant 125\,000$ (Wagstaff [42]); no example contradicting it is known; but a general proof is lacking.

The best general result follows from the solution in 1983 of the Mordell Conjecture by Gerd Faltings; this implies that for any $n \geqslant 3$ there can be *at most a finite number* of (pairwise coprime) solutions.

This means that Fermat's 'Last Theorem' is a misnomer – doubly so in fact – for not only is a proof unknown, which reduces its status from 'theorem' to 'conjecture'; it was also not the last of Fermat's many results. It was simply the last of his problems left to posterity which remained unsolved in the mid 19th century. Despite the objections, the name has stuck, and with it is a glow of romanticism which is lacking in the more accurate title 'the Fermat Conjecture'. It has the two classic ingredients of a problem to catch the imagination of a wider public – a simple statement that can be widely understood but a proof which defeats the greatest intellects.

Another classic problem of this type – 'the impossibility of trisecting an angle using ruler and compasses alone' – took two thousand years to be solved. This was posed by the Greeks in studying geometry and was solved in the early 19th century using algebraic techniques. In the same way the advancement in the solution of Fermat's Last Theorem has moved out of the original domain, the theory of natural numbers, to a different area of mathematical study, algebraic numbers. In the 19th century the developing theory of algebra had matured to a state where it could usefully be applied in number theory.

As it happened, Fermat's Last Theorem was not the major problem being attacked by number theorists at the time; for example, when Kummer made the all-important breakthrough that we are to describe in this text, he was working on a different problem, a topic called the 'higher reciprocity laws'. At this stage it is worth making a minor diversion to look at this subject, for it was here that algebraic numbers entered number theory in the work of Gauss. As an eighteen-year-old, in 1796 Gauss had given the first proof of a remarkable fact observed empirically by Euler in 1783. Euler had addressed himself to the problem of when an integer q was congruent to a perfect square modulo a prime p,

$$x^2 \equiv q \pmod{p}.$$

In such a case, q is said to be a *quadratic residue* of p. Euler concentrated on the case when p, q were distinct odd primes and noted: if at least one of the odd primes p, q is of the form $4r + 1$, then q is a quadratic residue of p if and only if p is a quadratic residue of q; on the other hand, if both p, q are of the form $4r + 3$, then precisely one is a quadratic residue of the other.

Because of the reciprocal nature of the relationship between p and q, this result was known as the *quadratic reciprocity law*. Legendre attempted a proof in 1785 but assumed that certain arithmetic progressions contained infinitely many primes – a theorem whose proof turned out to be far deeper than the quadratic reciprocity law itself. Legendre introduced the symbol

$$(q/p) = \begin{cases} 1 & \text{if } q \text{ is a quadratic residue of } p \\ -1 & \text{if not,} \end{cases}$$

in terms of which the law becomes

$$(q/p)(p/q) = (-1)^{(p-1)(q-1)/4}.$$

When Gauss gave the first proof in 1796 he was dissatisfied because his method did not seem a natural way to attack so seemingly simple a theorem. He went on to give several more proofs, two of which appeared in his book *Disquisitiones Arithmeticae* (1801), a definitive text on number theory which still remains in print [24]. His second proof depends on a numerical criterion that he discovered, and we give a computational proof depending on this criterion in the Appendix.

Between 1808 and 1832 Gauss continued to look for similar laws for powers higher than squares. This entailed looking for relationships between p and q so that q was a cubic residue of p ($x^3 \equiv q \pmod{p}$) or a biquadratic residue ($x^4 \equiv q \pmod{p}$), and so on. He found certain higher reciprocity laws, but in doing so he discovered that his calculations were made easier by working over the Gaussian integers $a + bi$ ($a, b \in \mathbf{Z}$, $i = \sqrt{-1}$), rather than the integers alone. He developed a theory of prime factorization for these, proved that decomposition into primes was unique, and developed a law of biquadratic reciprocity. In the same way he considered cubic reciprocity by using numbers of the form $a + b\omega$ where $\omega = e^{(2\pi i)/3}$. These higher reciprocity laws do not have the same striking simplicity as quadratic reciprocity and we shall not study them in this text. But Gauss' use of these new types of number is of fundamental importance in Fermat's Last Theorem, and the study of their factorization properties is a deep and fruitful source of methods and problems.

The numbers concerned are all examples of a particular type of complex number; namely, one which is a solution of a polynomial equation

$$a_n x^n + \ldots + a_1 x + a_0 = 0$$

where all the coefficients are integers. Such a complex number is said to be *algebraic*, and if $a_n = 1$, it is called an *algebraic integer*. Examples of algebraic integers include i (which satisfies $x^2 + 1 = 0$), $\sqrt{2}$ ($x^2 - 2 = 0$) and more complicated examples, such as the roots of $x^7 - 265x^3 + 7x^2 - 2x + 329 = 0$. The number $\frac{1}{2}i$ (satisfying $4x^2 + 1 = 0$) is algebraic but not an integer. On the other hand, there are complex numbers which are not algebraic, such as e or π.

In the wider setting of algebraic integers we can factorize a solution of Fermat's equation $x^n + y^n = z^n$ (if one exists) by introducing a complex nth root of unity, $\zeta = e^{2\pi i/n}$, and writing (2) as

$$(x + y)(x + \zeta y) \ldots (x + \zeta^{n-1} y) = z^n. \tag{3}$$

If $\mathbf{Z}[\zeta]$ denotes the set of algebraic integers of the form $a_0 + a_1\zeta + \ldots + a_r\zeta^r$ where each a_r is an ordinary integer, then this factorization takes place in the ring $\mathbf{Z}[\zeta]$.

In 1847 the French mathematician Lamé announced a 'proof' of Fermat's Last Theorem. In outline his proposal was to show that only the case where x, y have no common factors need be considered; and then deduce that in this case $x + y, x + \zeta y, \ldots, x + \zeta^{n-1} y$ have no common factors, that is, they are relatively prime. He then argued that a product of relatively prime numbers in (3) can be equal to an nth power only if each of the factors is an nth power. So

$$\begin{aligned} x + y &= u_1^n \\ x + \zeta y &= u_2^n \\ &\cdots\cdots \\ x + \zeta^{n-1} y &= u_n^n. \end{aligned} \tag{4}$$

On this basis Lamé derived a contradiction.

It was immediately pointed out to him by Liouville that the deduction of (4) from (3) assumed uniqueness of factorization in a very subtle way. Liouville's fears were confirmed when he later received a letter from Kummer who had shown that uniqueness of factorization fails in some cases, the first being $n = 23$. Over the summer of 1847 Kummer went on to devise his own proof of Fermat's Last Theorem for certain exponents n, surmounting the difficulties of non-uniqueness of factorization by introducing the theory of 'ideal' complex numbers. In retrospect this theory can be viewed as introducing numbers from outside $\mathbf{Z}[\zeta]$ to use as factors when factorizing elements within $\mathbf{Z}[\zeta]$. These 'ideal factors' restore a version of unique factorization.

Nowadays the theory takes a rather different form from that in which Kummer left it, but the key concept of an 'ideal' (a reformulation by Dedekind of Kummer's 'ideal number') remains. By using his theory of ideal numbers Kummer proved Fermat's Last Theorem for a wide range of prime numbers – the so-called 'regular' primes. He also evolved a powerful machine with applications to many other problems in mathematics. In fact a large part of classical number theory can be expressed in the framework of algebraic numbers. This point of view was urged most strongly by Hilbert in his *Zahlbericht* of 1897, which had an enormous influence on the development of number theory. As a result, algebraic number theory is today a flourishing and important branch of mathematics, with deep methods and insights; and – most significantly – applications not only to number theory but also to group theory, algebraic geometry, topology and analysis.

PART I

Algebraic methods

CHAPTER ONE

Algebraic background

In this introductory chapter we recall some fundamental facts about rings, fields, polynomials, modules, and abelian groups; to establish a firm basis for the rest of the book. Many of these results will be familiar, but we briefly give definitions to establish conventions regarding terminology. We expect the reader to be acquainted with the elementary properties of groups, rings and fields; and to have a knowledge of linear algebra over an arbitrary field up to simple properties of determinants. Familiar results at this level will be stated without proof; results which we consider may be less familiar to some readers may be proved in full or in outline as we consider appropriate. All results not proved in full have complete proofs in [HH] or [GT], to which references are given where relevant.

Our plan of attack is guided by the observation that most of the algebraic number theory in this book takes place in subfields and subrings of the complex numbers. First, therefore, we set up the ring theoretic language (and in particular the notion of an ideal, which proves to be so important). Then we consider factorization of polynomials over a ring (which in this book will often be a subfield of the complex numbers). Topics of central importance at this stage are the factorization of a polynomial over an extension field and the theory of elementary symmetric polynomials. Module-theoretic language will help us clarify certain points later; and results concerning finitely generated abelian groups are proved because they are vital in describing the properties of the additive group structure of the subrings of the complex numbers which occur. With the prologue of the first chapter behind us we shall then be ready to begin the main action.

1.1 Rings and fields

Unless explicitly stated to the contrary, the term *ring* in this book will always mean a commutative ring R with identity element 1 (or 1_R). If such a ring has no zero-divisors (so that in R, $a \neq 0$, $b \neq 0$ implies $ab \neq 0$), and if $1 \neq 0$ in it then it will be called a *domain* (or *integral domain*). An element a in a ring R is called a *unit* if there exists $b \in R$ such that $ab = 1$. Such an element b is, of course, unique and will be denoted by a^{-1}; ca^{-1} will also be denoted by c/a. If $1 \neq 0$ in R and every non-zero element in R is a unit, then R will be called a *field.*

We shall use the standard notation $\mathbf{N}$ for the set of natural numbers $\{0, 1, 2, \ldots\}$, $\mathbf{Z}$ for the integers (Zahlen), $\mathbf{Q}$ for the rationals (*q*uotients), $\mathbf{R}$ for the reals and $\mathbf{C}$ for the complex numbers. Under the usual operations $\mathbf{Q}, \mathbf{R}, \mathbf{C}$ are fields, $\mathbf{Z}$ is a domain and $\mathbf{N}$ is not even a ring. For $n \in \mathbf{N}$, $n > 0$, we denote the ring of integers modulo n by $\mathbf{Z}_n$. If n is composite, then $\mathbf{Z}_n$ has zero divisors, but for n prime, then $\mathbf{Z}_n$ is a field ([GT] Theorem 1.1, p. 3).

A subring S of a ring R will be required to contain 1_R. We can check that S is a subring by demonstrating that $1_R \in S$, and if $s, t \in S$ then $s + t, -s, st \in S$. It then forms a ring in its own right under the operations restricted from R. In the same way, if K is a field, then a subfield F of K is a field under the operations restricted from K. We can check that F is a subfield of K by demonstrating $1_K \in F$, and if $s, t \in F$ $(s \neq 0)$, then $s + t, -s, st, s^{-1} \in F$.

The concept of an *ideal* will be of central importance in this text. Recall that an ideal is a non-empty subset I of a ring

R such that if $r, s \in I$, then $r - s \in I$; and if $r \in R$, $s \in I$ then $rs \in I$. We shall also require the concept of the quotient ring R/I of R by an ideal I. The elements of R/I are cosets $I + r$ of the additive group of I in R, with addition and multiplication defined by

$$(I + r) + (I + s) = I + (r + s)$$
$$(I + r)(I + s) = I + rs$$

for all $r, s \in R$.

A *homomorphism* f from a ring R_1 to a ring R_2 is a function $f: R_1 \to R_2$ such that

$$f(1_{R_1}) = 1_{R_2}$$
$$f(r + s) = f(r) + f(s)$$
$$f(rs) = f(r)f(s)$$

for all $r, s \in R_1$. A *monomorphism* is an injective homomorphism and an *isomorphism* is a bijective homomorphism.

The *kernel* and *image* of a homomorphism f are defined in the usual way:

$$\ker f = \{r \in R_1 \mid f(r) = 0\}$$
$$\operatorname{im} f = \{f(r) \in R_2 \mid r \in R_1\}.$$

The kernel is an ideal of R_1; the image is a subring of R_2; and there is an isomorphism from $R_1/\ker f$ to $\operatorname{im} f$. (Students requiring explanations of the relevant theory may consult [HH] pp. 18–26.)

If X and Y are subsets of a ring R we write $X + Y$ for the set of all elements $x + y$ ($x \in X$, $y \in Y$); and XY for the set of all finite sums $\Sigma\, x_i y_i$ ($x_i \in X$, $y_i \in Y$). When X and Y are both ideals, so are $X + Y$ and XY.

The sum $X + Y$ of two sets can be generalized to an arbitrary collection $\{X_i\}_{i \in I}$ by defining $\Sigma_{i \in I} X_i$ to be the set of all finite sums $x_{i_1} + \ldots + x_{i_n}$ of elements $x_{i_j} \in X_{i_j}$.

We shall make the customary abuses of notation with regard to $\{x\}$ and x, writing for example xY for $\{x\}Y$, $x + Y$ for $\{x\} + Y$, and 0 for $\{0\}$.

The ideal *generated* by a subset X of R is the smallest ideal of R containing X; we shall denote this by $\langle X \rangle$. If $X = \{x_1, \ldots, x_n\}$, then we shall write $\langle X \rangle$ as $\langle x_1, \ldots, x_n \rangle$. (Some writers use (X) where we have written $\langle X \rangle$, but then the last-mentioned simplification of notation would reduce to the notation for an n-tuple $(x_1, \ldots, x_n)$, so $\langle X \rangle$ is to be preferred.)

A simple calculation shows that†

$$\langle X \rangle = XR = \sum_{x \in X} xR.$$

If there exists a finite subset $X = \{x_1, \ldots, x_n\}$ of R such that $I = \langle X \rangle$, then we say that I is *finitely generated* as an ideal of R. If $I = \langle x \rangle$ for an element $x \in R$ we say that I is the *principal ideal* generated by x.

Example 1. Let $R = \mathbf{Z}$, $X = \{4, 6\}$, then $\langle 4, 6 \rangle$ is finitely generated. In fact $\langle 4, 6 \rangle$ contains $2.4 - 6 = 2$ and it easily follows that $\langle 4, 6 \rangle = \langle 2 \rangle$, so that further this ideal is principal. More generally, every ideal of $\mathbf{Z}$ is of the form $\langle n \rangle$ for some $n \in N$, hence principal.

Example 2. Let R be the set $\mathbf{Q}$ under the usual operation of addition, but define a non-standard multiplication on R by setting $xy = 0$ for all $x, y \in R$. The ideal $\langle X \rangle$ for a subset $X \subseteq R$ is then equal to the abelian group generated by X under addition. Now R is an ideal of R, but is not finitely generated. To see this, suppose that R is generated as an abelian group by elements $p_1/q_1, \ldots, p_n/q_n$. Then the only primes dividing the denominators of elements of R will be those dividing $q_1, \ldots, q_n$, which is a contradiction.

If K is a field and R is a subring of K then R is a domain. Conversely, every domain D can be embedded in a field L; and there exists such an L consisting only of elements d/e

† The identity element 1_R is crucial in this equation. In a commutative ring without identity we would also have to add on the additive group generated by X to XR and to $\Sigma_{x \in X}\, xR$.

where $d, e \in D$ and $e \neq 0$. Such an L, which is unique up to isomorphism, is called the *field of fractions* of D. (See [HH] pp. 50–51.)

Theorem 1.1. *Every finite integral domain is a field.*

Proof. Let D be a finite integral domain. Since $1 \neq 0$, then D has at least 2 elements. For $0 \neq x \in D$ the elements xy, as y runs through D, are distinct; for if $xy = xz$ then $x(y - z) = 0$ and so $y = z$ since D has no zero-divisors. Hence, by counting, the set of all elements xy is D. Thus $1 = xy$ for some $y \in D$, and therefore D is a field. □

Every field has a unique minimal subfield, the *prime subfield,* and this is isomorphic either to $\mathbf{Q}$ or to $\mathbf{Z}_p$ where p is a prime number. (See [GT], Theorem 1.2, p. 4.) Correspondingly, we say that the *characteristic* of the field is 0 or p. In a field of characteristic p we have $px = 0$ for every element x, where as usual we write

$$px = (1 + 1 + \ldots + 1)x$$

where there are p summands 1; and p is the smallest positive integer with this property. In a field of characteristic zero, if $nx = 0$ for some non-zero element x and integer n, then $n = 0$. Our major concern in the sequel will be subfields of $\mathbf{C}$ (the complex numbers), which of course have characteristic zero; but fields of prime characteristic will arise naturally from time to time.

We shall use without further comment the fact that $\mathbf{C}$ is *algebraically closed*: given any polynomial p over $\mathbf{C}$ there exists $x \in \mathbf{C}$ such that $p(x) = 0$. For a proof of this see [GT] pp. 193 ff.; different proofs using analysis or topological considerations are in Hardy [27] p. 492 and Titchmarsh [41] p. 118.

1.2 Factorization of polynomials

Later in the book we shall consider factorization in a more general context. Here we concentrate on factorizing polynomials. First a few general remarks.

In a ring S, if we can write $a = bc$ for $a, b, c \in S$, then we say that b, c are *factors* of a. We also say 'b divides a' and write

$$b \mid a.$$

For any unit $e \in S$ we can always write

$$a = e(e^{-1}a),$$

so, trivially, a unit is a factor of all elements in S. If $a = bc$ where neither b nor c is a unit, then b and c are called *proper factors* and a is said to be *reducible.* In particular, $0 = 0.0$ is reducible.

Note that if a is itself a unit and $a = bc$, we have

$$1 = aa^{-1} = bca^{-1},$$

so b and c are both units. A unit cannot have a proper factorization. We therefore concentrate on factorization of non-units. A non-unit $a \in S$ is said to be *irreducible* if it has no proper factors.

Now we turn our attention to the case $S = R[t]$, the ring of polynomials in an indeterminate t with coefficients in a ring R. The elements of $R[t]$ are expressions

$$r_n t^n + r_{n-1} t^{n-1} + \ldots + r_1 t + r_0$$

where $r_0, r_1, \ldots, r_n \in R$ and addition and multiplication are defined in the obvious way. (For a formal treatment of polynomials see [HH] p. 37.)

Given a non-zero polynomial

$$p = r_n t^n + \ldots + r_0,$$

we define the *degree* of p to be the largest value of n for which $r_n \neq 0$, and write it ∂p. Polynomials of degree 0, 1, 2, 3, 4, 5, . . . , will often be referred to as *constant, linear, quadratic, cubic, quartic, quintic, . . .*, polynomials respectively. In particular a constant polynomial is just a (non-zero) element of R.

If R is an integral domain, then

$$\partial pq = \partial p + \partial q$$

for non-zero p, q so $R[t]$ is also an integral domain. If $p = aq$ in $R[t]$, then $\partial p = \partial a + \partial q$ implies that

$$\partial q \leqslant \partial p.$$

When R is not a field, then it is perfectly possible to have a non-trivial factorization in which $\partial p = \partial q$. For example

$$3t^2 + 6 = 3(t^2 + 2)$$

in $\mathbf{Z}[t]$, where neither 3 nor $t^2 + 2$ is a unit. This is because of the existence of non-units in R. However, if R is a field, then all (non-zero) constants in $R[t]$ are units and so if q is a proper factor of p for polynomials over a field, then $\partial q < \partial p$.

Let us concentrate for a time on polynomials over a field K. Here we have the *division algorithm* which states that if $p, q \neq 0$ then

$$p = qs + r$$

where either $r = 0$ or $\partial r < \partial q$. The proof is by induction on ∂p and in practice is no more than long division of p by q leaving remainder r, which is either zero (in which case $q \mid p$) or of degree lower than q.

The division algorithm is used repeatedly in the *Euclidean algorithm,* which is a particularly efficient method for finding the *highest common factor* d of non-zero polynomials p, q. This is defined by the properties:

(a) $d \mid p, d \mid q,$

(b) If $d' \mid p$ and $d' \mid q$ then $d' \mid d$.

These define d uniquely up to non-zero constant multiples. To calculate d we first suppose that p, q are named so that $\partial p \geqslant \partial q$; then divide q into p to get

$$p = qs_1 + r_1 \qquad \partial r_1 < \partial q \leqslant \partial p,$$

and continue in the following way:

$$\begin{aligned} q &= r_1 s_2 + r_2 & \partial r_2 &< \partial r_1 \\ r_1 &= r_2 s_3 + r_3 & \partial r_3 &< \partial r_2 \\ &\cdots\cdots\cdots\cdots\cdots & & \\ r_{n-2} &= r_{n-1} s_n + r_n & \partial r_n &< \partial r_{n-1} \end{aligned}$$

until we arrive at a zero remainder:

$$r_{n-1} = r_n s_{n+1}.$$

The last non-zero remainder r_n is the highest common factor. (From the last equation $r_n \mid r_{n-1}$, and working back successively, r_n is a factor of $r_{n-2}, \ldots, r_1, p, q$, verifying (a). If $d' \mid p, d' \mid q$, then from the first equation, d' is a factor of $r_1 = p - qs_1$, and successively working down the equations, d' is a factor of $r_2, r_3, \ldots, r_n$, so $d' \mid r_n$, verifying (b).) Beginning with the first equation, and substituting in those which follow, we find that $r_i = a_i p + b_i q$ for suitable $a_i, b_i \in K[t]$, and in particular the highest common factor $d = r_n$ is of the form

$$d = ap + bq \qquad \text{for suitable } a, b \in K[t]. \tag{1}$$

A useful special case is when $d = 1$, when p, q are called *coprime* and (1) gives

$$ap + bq = 1 \qquad \text{for suitable } a, b \in K[t].$$

This technique for calculating the highest common factor can also be used to find the polynomials a, b.

Example. $p = t^3 + 1, q = t^2 + 1 \in \mathbb{Q}[t]$.
Then

$$\begin{aligned} t^3 + 1 &= t(t^2 + 1) + (-t + 1), \\ t^2 + 1 &= (-t - 1)(-t + 1) + 2, \\ -t + 1 &= (-\tfrac{1}{2}t + \tfrac{1}{2})2. \end{aligned}$$

The highest common factor is 2, or up to a constant factor, 1, so p and q are coprime, and substituting back from the second equation,

$$1 = \tfrac{1}{2}(t^2+1) + \tfrac{1}{2}(t+1)(-t+1).$$

Then substituting for $-t+1$ using the first equation:

$$\begin{aligned} 1 &= \tfrac{1}{2}(t^2+1) + \tfrac{1}{2}(t+1)((t^3+1) - t(t^2+1)) \\ &= (-\tfrac{1}{2}t^2 - \tfrac{1}{2}t + \tfrac{1}{2})(t^2+1) + (\tfrac{1}{2}t + \tfrac{1}{2})(t^3+1). \end{aligned}$$

Factorizing a single polynomial p is by no means as straightforward as finding the highest common factor of two. It is known that every non-zero polynomial over a field K is a product of finitely many irreducible factors, and these are unique up to the order in which they are multiplied and up to constant factors. (See [HH] pp. 60–1 or [GT] pp. 19–21.) Finding these factors is very much an *ad hoc* matter. Linear factors are easiest, since $(x-\alpha) \mid p$ if and only if $p(\alpha) = 0$.

If $p(\alpha) = 0$, then α is called a *zero* of p. If $(t-\alpha)^m \mid p$ where $m \geqslant 2$, then α is a *repeated zero* and the largest such m is the *multiplicity* of α.

To detect repeated zeroes, we use a method which was (like much in this chapter) far more familiar at the turn of the century than now. Given a polynomial

$$f = \sum_{i=0}^{n} r_i t^i$$

over a ring R we define

$$\mathrm{D}f = \sum_{i=0}^{n} i r_i t^{i-1},$$

called for obvious reasons the *formal derivative* of f. It is not hard to check directly that

$$\begin{aligned} \mathrm{D}(f+g) &= \mathrm{D}f + \mathrm{D}g \\ \mathrm{D}(fg) &= (\mathrm{D}f)g + f(\mathrm{D}g). \end{aligned}$$

This enables us to check for repeated factors. A factor q of a polynomial p is *repeated* if $q^r \mid p$ for some $r \geqslant 2$. In particular q is repeated if its square divides p.

Theorem 1.2. *Let K be a field of characteristic zero. A non-zero polynomial f over K is divisible by the square of a polynomial of degree > 0 if and only if f and $\mathrm{D}f$ have a common factor of degree > 0*

Proof. First suppose $f = g^2 h$. Then

$$\mathrm{D}f = g^2 \mathrm{D}h + 2g(\mathrm{D}g)h$$

and so f and $\mathrm{D}f$ have g as a common factor.

Now suppose that f has no squared irreducible factor. Then for any irreducible factor g of f we find

$$f = gh$$

where g and h are coprime (otherwise g would be a factor of h and would occur as a squared factor in gh). Were f and $\mathrm{D}f$ to have a common factor g, which we take to be irreducible, then on differentiating formally we would obtain

$$\mathrm{D}f = (\mathrm{D}g)h + g(\mathrm{D}h).$$

So g is a factor of $(\mathrm{D}g)h$, hence of $\mathrm{D}g$ because g and h are coprime. But $\mathrm{D}g$ is of lower degree than g, hence it can only have g as a factor if $\mathrm{D}g = 0$. Since K has characteristic zero, by direct computation, this implies g is constant, so f and g can have no non-trivial common factor. □

Remark. If the field has characteristic $p > 0$, then the first part of Theorem 1.2, that f having a squared factor implies f and $\mathrm{D}f$ have a common factor, is still true, and the proof is the same as above.

A result which we shall need later is:

Corollary 1.3. *An irreducible polynomial over a subfield K of $\mathbf{C}$ has no repeated zeros in $\mathbf{C}$.*

Proof. Suppose f is irreducible over K. Then f and $\mathrm{D}f$ must be coprime (for a common factor would be a squared factor of f by 1.2, and f is irreducible). Thus there exist polynomials

a, b over K such that $af + b\mathrm{D}f = 1$, and the same equation interpreted over **C** shows f and $\mathrm{D}f$ to be coprime over **C**. By Theorem 1.2 again, f cannot have repeated zeros. □

We shall often consider factorization of polynomials over **Q**. When such a polynomial has integer coefficients we shall find that we need consider only factors which themselves have integer coefficients. This fact is enshrined in a result due to Gauss:

Lemma 1.4. *Let* $p \in \mathbf{Z}[t]$, *and suppose that* $p = gh$ *where* $g, h \in \mathbf{Q}[t]$. *Then there exists* $\lambda \in \mathbf{Q}$, $\lambda \neq 0$, *such that* λg, $\lambda^{-1}h \in \mathbf{Z}[t]$.

Proof. Multiplying by the product of the denominators of the coefficients of g, h we can rewrite $p = gh$ as

$$np = g'h'$$

where g', h' are rational multiples of g, h respectively, $n \in \mathbf{Z}$ and $g', h' \in \mathbf{Z}[t]$. This means that n divides the coefficients of the product $g'h'$. We now divide the equation successively by the prime factors of n. We shall establish that if k is a prime factor of n, then k divides all the coefficients of g' or all those of h'. Whichever it is, we can divide that particular polynomial by k to give another polynomial with integer coefficients. After dividing in this way by all the prime factors of n, we are left with

$$p = \bar{g}\bar{h}$$

where $\bar{g}, \bar{h} \in \mathbf{Z}[t]$ are rational multiples of g, h respectively. Putting $\bar{g} = \lambda g$ for $\lambda \in \mathbf{Q}$, we obtain $\bar{h} = \lambda^{-1}h$ and the result will follow.

It remains to prove that if

$$g' = g_0 + g_1 + \ldots + g_r t^r$$
$$h' = h_0 + h_1 + \ldots + g_s t^s$$

and a prime k divides all the coefficients of $g'h'$, then k must divide all the g_i or all the h_j. But if a prime k does not divide all the g_i and all the h_j, we can choose the *first* of each set of coefficients, say g_m, h_q which are not divisible by k. Then the coefficient of t^{m+q} in the product $g'h'$ is

$$g_0 h_{m+q} + g_1 h_{m+q-1} + \ldots + g_m h_q + \ldots g_{m+q} h_0$$

and since every term in this expression is divisible by k except $h_q g_m$, this would mean that the whole coefficient would not be divisible by k, a contradiction. □

We shall need methods for proving irreducibility of various specific polynomials over **Z**. The first of these is known as *Eisenstein's criterion*:

Theorem 1.5. *Let*

$$f = a_0 + a_1 t + \ldots + a_n t^n$$

be a polynomial over **Z**. *Suppose there is a prime q such that*

(a) $q \nmid a_n$,
(b) $q \mid a_i \qquad (i = 0, 1, \ldots, n-1)$,
(c) $q^2 \nmid a_0$.

Then, apart from constant factors, f is irreducible over **Z**, *and hence irreducible over* **Q**.

Proof. By Lemma 1.4 it is enough to show that f can only have constant factors over **Z**.

If not, then $f = gh$ where

$$g = g_0 + g_1 t + \ldots + g_r t^r$$
$$h = h_0 + h_1 t + \ldots + h_s t^s$$

with all $g_i, h_j \in$ **Z** and $r, s > 1$, $r + s = n$.

Now $g_0 h_0 = a_0$, so (b) implies q divides one of g_0, h_0 whilst (c) implies it cannot divide both. Without loss in generality, suppose q divides g_0 but not h_0. Not all g_i are divisible by q because this would imply that q divides a_n, contrary to (a). Let g_m be the first coefficient of g not divisible by q. Then

$$a_m = g_0 h_m + \ldots + g_m h_0$$

where $m \leqslant r < n$. All the summands on the right are divisible by q except the last, which means that a_m is not divisible by q, contradicting (b). □

A second useful method is *reduction modulo n*, as follows. Suppose $0 \neq p \in \mathbf{Z}[t]$, with p reducible: say $p = qr$. The natural homomorphism $\mathbf{Z} \to \mathbf{Z}_n$ gives rise to a homomorphism $\mathbf{Z}[t] \to \mathbf{Z}_n[t]$. Using bars to denote images under this map, we have $\bar{p} = \bar{q}\bar{r}$. If $\partial \bar{p} = \partial p$, then clearly $\partial \bar{q} = \partial q$, $\partial \bar{r} = \partial r$, and $\bar{p}$ is also reducible. This proves:

Theorem 1.6. *If $p \in \mathbf{Z}[t]$ and its image $\bar{p} \in \mathbf{Z}_n[t]$ is reducible, with $\partial \bar{p} = \partial p$, then p is irreducible as an element of $\mathbf{Z}_n[t]$.* □

In practice we take n to be prime, though this is not essential. The point of reducing modulo n is that $\mathbf{Z}_n$ being finite, there are only a finite number of possible factors of $\bar{p}$ to be considered.

Examples.

1 The polynomial $t^2 - 2$ satisfies Eisenstein's criterion with $q = 2$.

2 The polynomial $t^{11} - 7t^6 + 21t^5 + 49t - 56$ satisfies Eisenstein's criterion with $q = 7$.

3 The polynomial $t^5 - t + 1$ does not satisfy Eisenstein's criterion for any q. Instead we try reduction modulo 5. There is no linear factor since none of 0, 1, 2, 3, 4 yield 0 when substituted for t, so the only possible way to factorize is

$$t^5 - t + 1 = (t^2 + \alpha t + \beta)(t^3 + \gamma t^2 + \delta t + \epsilon)$$

where $\alpha, \beta, \gamma, \delta, \epsilon$ take values 0, 1, 2, 3 or 4 (mod 5). This gives a system of equations on comparing coefficients: there are only a finite number of possibilities all of which are easily eliminated. Hence the polynomial is irreducible mod 5, so irreducible over $\mathbf{Z}$.

1.3 Field extensions

In finding the zeros of a polynomial p over a field K it is often necessary to pass to a larger field L containing K. In these circumstances, L is called a *field extension* of K. For example, $p(t) = t^2 + 1$ has no zeros in $\mathbf{R}$, but considering p as a polynomial over $\mathbf{C}$, it has zeros $\pm i$ and a factorization

$$p(t) = (t+i)(t-i).$$

Field extensions often arise in a slightly more general context as a monomorphism $j: K \to L$ where K and L are fields. We shall see such instances shortly. It is customary in these cases (see [GT] p. 33) to identify K with its image $j(K)$, which is a subfield of L; then a field extension is a pair of fields (K, L) where K is a subfield of L. We talk of the extension

$$L:K$$

of K. Most field extensions with which we shall deal will involve two subfields of $\mathbf{C}$.

If $L:K$ is a field extension, then L has a natural structure as a vector space over K (where vector addition is addition in L and scalar multiplication of $\lambda \in K$ on $v \in L$ is just $\lambda v \in L$). The dimension of this vector space is called the *degree* of the extension, or the *degree of L over K*, and written

$$[L:K].$$

The degree has an important multiplicative property:

Theorem 1.7. *If $H \subseteq K \subseteq L$ are fields, then*

$$[L:H] = [L:K][K:H].$$

Proof. We sketch this. Details are in [GT], Theorem 4.2 p. 50. Let $\{a_i\}_{i \in I}$ be a basis for L over K, and $\{b_j\}_{j \in J}$ a basis for K over H. Then $\{a_i b_j\}_{(i,j) \in I \times J}$ is a basis for L over H. □

If $[L:K]$ is finite we say that L is a *finite extension* of K. Given a field extension $L:K$ and an element $\alpha \in L$, there may or may not exist a polynomial $p \in K[t]$ such that $p(\alpha) = 0, p \neq 0$. If not, we say that α is *transcendental* over K. If such a p exists, we say that α is *algebraic* over K. If α is algebraic over K, then there exists a unique monic polynomial q of minimal degree subject to $q(\alpha) = 0$, and q is called the *minimum polynomial* of α over K. (A *monic* polynomial is one with highest coefficient 1.) The minimum polynomial of α is irreducible over K. (These facts are to be found in [GT] pp. 38, 39.)

If $\alpha_1, \ldots \alpha_n \in L$, we write

$$K(\alpha_1, \ldots, \alpha_n)$$

for the smallest subfield of L containing K and the elements $\alpha_1, \ldots, \alpha_n$.

In an analogous way, if S is a subring of a ring R and $\alpha_1, \ldots, \alpha_n \in R$, we write

$$S[\alpha_1, \ldots, \alpha_n]$$

for the smallest subring of R containing S and the elements $\alpha_1, \ldots, \alpha_n$. Clearly $S[\alpha_1, \ldots, \alpha_n]$ consists of all polynomials in $\alpha_1, \ldots, \alpha_n$ with coefficients in S. For instance $S[\alpha]$ consists of polynomials

$$s_0 + s_1\alpha + \ldots + s_m\alpha^m \qquad (s_i \in S).$$

The case of $K(\alpha)$ is more interesting. If α is transcendental over K, then for $k_m \neq 0$ we have

$$k_0 + k_1\alpha + \ldots + k_m\alpha^m \neq 0 \qquad (k_i \in K).$$

In this case $K(\alpha)$ must include all rational expressions

$$\frac{s_0 + s_1\alpha + \ldots + s_n\alpha^n}{k_0 + k_1\alpha + \ldots + k_m\alpha^m} \qquad (s_j, k_i \in K, k_m \neq 0)$$

and clearly consists precisely of these elements.

However, for α algebraic, we have:

Theorem 1.8. *If $L:K$ is a field extension and $\alpha \in L$, then α is algebraic over K if and only if $K(\alpha)$ is a finite extension of K. In this case, $[K(\alpha):K] = \partial p$ where p is the minimum polynomial of α over K, and $K(\alpha) = K[\alpha]$.*

Proof. Once more we sketch the proof, given in full in [GT] proposition 4.3, p. 52. If $[K(\alpha):K] = n < \infty$ then the powers, $1, \alpha, \alpha^2, \ldots, \alpha^n$ are linearly dependent over K, whence α is algebraic. Conversely, suppose α algebraic with minimum polynomial p of degree m. We claim that $K(\alpha)$ is the vector space over K spanned by $1, \alpha, \ldots, \alpha^{m-1}$. This (call it V) is certainly closed under addition, subtraction, and multiplication by α; for the last statement note that $\alpha^m = -p(\alpha) + \alpha^m = q(\alpha)$ where $\partial q < m$. Hence V is closed under multiplication, and so forms a ring. All we need prove now is that if $0 \neq v \in V$ then $1/v \in V$. Now $v = h(\alpha)$ where $h \in K[t]$ and $\partial h < m$. Since p is irreducible, p and h are coprime, so there exist $f, g \in K[t]$ such that

$$f(t)p(t) + g(t)h(t) = 1.$$

Then

$$1 = f(\alpha)p(\alpha) + g(\alpha)h(\alpha) = g(\alpha)h(\alpha)$$

so that $1/v = g(\alpha) \in V$ as required. But it follows at once that $[K(\alpha):K] = \dim_K V = m$. □

If we specify in advance K and an irreducible monic polynomial $p(t) \in K[t]$ then there exists up to isomorphism a unique extension field L such that L contains an element α with minimum polynomial p, and $L = K(\alpha)$. This can be constructed as $K[t]/\langle p \rangle$. It is customary to express this construction by the phrase 'adjoin to K an element α with $p(\alpha) = 0$' and to write $K(\alpha)$ for the resulting field. This, and much else, is discussed in [GT] Chapter 3, pp. 33–45.

1.4 Symmetric polynomials

Let $R[t_1, t_2, \ldots, t_n]$ denote the ring of polynomials in indeterminates $t_1, t_2, \ldots, t_n$ with coefficients in a ring R. Let S_n denote the symmetric group of permutations on $\{1, 2, \ldots, n\}$. For any permutation $\pi \in S_n$ and any polynomial $f \in R[t_1, \ldots, t_n]$ we define the polynomial f^π by

$$f^\pi(t_1, \ldots, t_n) = f(t_{\pi(1)}, \ldots, t_{\pi(n)}).$$

For example if $f = t_1 + t_2t_3$ and π is the cycle (123) then $f^\pi = t_2 + t_3t_1$. The polynomial f is *symmetric* if $f^\pi = f$ for all $\pi \in S_n$. For example $t_1 + \ldots + t_n$ is symmetric. More generally we have the *elementary symmetric polynomials*

$$s_r(t_1, \ldots, t_n) \qquad (1 \leqslant r \leqslant n)$$

defined to be the sum of all possible distinct products of r distinct t_i's. Thus

$$s_1(t_1, \ldots, t_n) = t_1 + t_2 + \ldots + t_n,$$
$$s_2(t_1, \ldots, t_n) = t_1t_2 + t_1t_3 + \ldots + t_2t_3 + \ldots + t_{n-1}t_n.$$
$$\ldots\ldots\ldots\ldots\ldots\ldots\ldots\ldots$$
$$s_n(t_1, \ldots, t_n) = t_1t_2 \ldots t_n.$$

These arise in the following circumstances: consider a polynomial of degree n over a subfield K of $\mathbf{C}$,

$$f = a_nt^n + \ldots + a_0,$$

and resolve it into linear factors over $\mathbf{C}$:

$$f = a_n(t - \alpha_1) \ldots (t - \alpha_n).$$

Then, expanding the product, we find

$$f = a_n(t^n - s_1t^{n-1} + \ldots + (-1)^n s_n),$$

where s_i denotes $s_i(\alpha_1, \ldots, \alpha_n)$.

A polynomial in $s_1, \ldots, s_n$ can clearly be rewritten as a symmetric polynomial in $t_1, \ldots, t_n$. The converse is also true, a fact first proved by Newton:

Theorem 1.9. *Let R be a ring. Then every symmetric polynomial in $R[t_1, \ldots, t_n]$ is expressible as a polynomial with coefficients in R in the elementary symmetric polynomials $s_1, \ldots, s_n$.*

Proof. We shall demonstrate a specific technique for reducing a symmetric polynomial into elementary ones. First we order the monomials $t_1^{\alpha_1} \ldots t_n^{\alpha_n}$ by a 'lexicographic' order in which $t_1^{\alpha_1} \ldots t_n^{\alpha_n}$ precedes $t_1^{\beta_1} \ldots t_n^{\beta_n}$ if the first non-vanishing

$\alpha_i - \beta_i$ is positive. Then given a polynomial $p \in R[t_1, \ldots, t_n]$, we order its terms lexicographically. If p is symmetric, then for every monomial $at_1^{\alpha_1} \ldots t_n^{\alpha_n}$ occurring in p, there occurs a similar monomial with the exponents permuted. Let α_1 be the highest exponent occurring in monomials of p: then there is a term containing $t_1^{\alpha_1}$. The leading term of p in lexicographic ordering contains $t_1^{\alpha_1}$, and among all such monomials we select the one with the highest occurring power of t_2 and so on. In particular, the leading term of a symmetric polynomial is of the form $at_1^{\alpha_1} \ldots t_n^{\alpha_n}$ where $\alpha_1 \geqslant \ldots \geqslant \alpha_n$. For example, the leading term of

$$s_1^{k_1} \ldots s_n^{k_n} = (t_1 + \ldots + t_n)^{k_1} \ldots (t_1 \ldots t_n)^{k_n}$$

is

$$t_1^{k_1 + \ldots + k_n} t_2^{k_2 + \ldots + k_n} \ldots t_n^{k_n}.$$

By choosing $k_1 = \alpha_1 - \alpha_2, \ldots, k_{n-1} = \alpha_{n-1} - \alpha_n$, $k_n = \alpha_n$ (which is possible because $\alpha_1 \geqslant \ldots \geqslant \alpha_n$), we can make this the same as the leading term of p. Then

$$p - as_1^{\alpha_1 - \alpha_2} \ldots s_{n-1}^{\alpha_{n-1} - \alpha_n} s_n^{\alpha_n}$$

has a lexicographic leading term

$$bt_1^{\beta_1} \ldots t_n^{\beta_n} \qquad (\beta_1 \geqslant \ldots \geqslant \beta_n)$$

which comes after $at_1^{\alpha_1} \ldots t_n^{\alpha_n}$ in the ordering. But only a finite number of monomials $t_1^{\gamma_1} \ldots t_n^{\gamma_n}$ satisfying $\gamma_1 \geqslant \ldots \geqslant \gamma_n$ follow $t_1^{\alpha_1} \ldots t_n^{\alpha_n}$ lexicographically, and so a finite number of repetitions of the given process reduce p to a polynomial in $s_1, \ldots, s_n$. □

Example. The symmetric polynomial

$$p = t_1^2 t_2 + t_1^2 t_3 + t_1 t_2^2 + t_1 t_3^2 + t_2^2 t_3 + t_2 t_3^2$$

is written lexicographically. Here $n = 3$, $\alpha_1 = 2$, $\alpha_2 = 1$, $\alpha_3 = 0$ and the method tells us to consider

$$p - s_1 s_2.$$

This simplifies to give

$$p - s_1 s_2 = 3t_1 t_2 t_3.$$

The polynomial $3t_1t_2t_3$ is visibly $3s_3$, but the method, using $\alpha_1 = \alpha_2 = \alpha_3 = 1$, also leads us to this conclusion.

This result about symmetric functions proves to be extremely useful in the following instance:

Corollary 1.10. *Suppose that L is an extension of the field K, $p \in K[t]$, $\partial p = n$ and the zeros of p are $\theta_1, \ldots, \theta_n \in L$. If $h(t_1, \ldots, t_n) \in K[t_1, \ldots, t_n]$ is symmetric, then $h(\theta_1, \ldots, \theta_n) \in K$.*

1.5 Modules

Let R be a ring. By an *R-module* we mean an abelian group M (written additively), together with a function $\alpha: R \times M \to M$, for which we write $\alpha(r, m) = rm$ ($r \in R$, $m \in M$), satisfying

(a) $(r + s)m = rm + sm$,
(b) $r(m + n) = rm + rn$,
(c) $r(sm) = (rs)m$
(d) $1m = m$

for all $r, s \in R$, $m, n \in M$.
(Although (d) is always obligatory in this text, be warned that in other parts of mathematics it may not be required to be so.) The function α is called an *R-action* on M.

If R is a field K, then an R-module is the same thing as a vector space over K. In this sense one may think of an R-module as a generalization of a vector space; but because of the lack of division in R, many of the techniques in vector space theory do not carry over as they stand to R-modules. The basic theory of modules may be found in [HH]. In particular we define an *R-submodule* of M to be a subgroup N of M (under addition) such that if $n \in N$, $r \in R$, then $rn \in N$. We may then define the *quotient module* M/N to be the corresponding quotient group, with R-action

$$r(N + m) = N + rm \qquad (r \in R,\ m \in M).$$

If $X \subseteq M$, $Y \subseteq R$, we define YX to be the set of all finite sums $\Sigma_i y_i x_i$ where $y_i \in Y$, $x_i \in X$.

The submodule of M *generated by* X, which we write

$$\langle X \rangle_R,$$

is the smallest submodule containing X. This is equal to RX. If $N = \langle x_1, \ldots, x_n \rangle_R$ then we say that N is a *finitely generated* R-module.

A $\mathbf{Z}$-module is nothing more than an abelian group M (written additively), and conversely, given an additive abelian group M we can make it into a $\mathbf{Z}$-module by defining

$$0m = 0, \qquad 1m = m \qquad (m \in M)$$

then inductively

$$(n+1)m = nm + m \qquad (n \in \mathbf{Z}, n > 0)$$

and

$$(-n)m = -nm \qquad (n \in \mathbf{Z}, n > 0).$$

We shall discuss this case further in the next section.

More generally there are several natural ways in which R-modules can arise, of which we distinguish three:

1. Suppose R is a subring of a ring S. Then S is an R-module with action

$$\alpha(r, s) = rs \qquad (r \in R, s \in S)$$

where the product is just that of elements in S.

2. Suppose I is an ideal of the ring R. Then I is an R-module under

$$\alpha(r, i) = ri \qquad (r \in R, i \in I)$$

where the product is that in R.

3. Suppose $J \subseteq I$ is another ideal: then J is also an R-module. The quotient module I/J has the action

$$r(J+i) = J + ri \qquad (r \in R, i \in I).$$

1.6 Free abelian groups

The study of algebraic numbers in this text will be carried out not only in subfields of $\mathbf{C}$, but also will require properties of subrings of **C**. A typical instance might be the subring

$$\mathbf{Z}[i] = \{a + ib \in \mathbf{C} \mid a, b \in \mathbf{Z}\}.$$

Considering the additive group of $\mathbf{Z}[i]$, we find that it is isomorphic to $\mathbf{Z} \times \mathbf{Z}$. More generally the additive group of those subrings of **C** that we shall study will usually be isomorphic to the direct product of a finite number of copies of **Z**. In this section we study the properties of such abelian groups which will prove of use later in this text.

Let G be an abelian group. In this section we shall use additive notation for G, so the group operation will be denoted by '+' the identity by 0, the inverse of g by $-g$ and powers of g by $2g, 3g, \ldots$. In later chapters we shall encounter cases where multiplicative notation is more appropriate and expect the reader to make the transition without undue fuss.

If G is finitely generated as a **Z**-module, so that there exist $g_1, \ldots, g_n \in G$ such that every $g \in G$ is a sum

$$g = m_1 g_1 + \ldots + m_n g_n \qquad (m_i \in \mathbf{Z})$$

then G is called a *finitely generated abelian group*.

Generalizing the notion of linear independence in a vector space, we say that elements $g_1, \ldots, g_n$ in an abelian group G are *linearly independent* (over **Z**) if any equation

$$m_1 g_1 + \ldots + m_n g_n = 0$$

with $m_1, \ldots, m_n \in \mathbf{Z}$ implies $m_1 = \ldots = m_n = 0$. A linearly independent set which generates G is called a *basis* (or a **Z**-basis for emphasis). If $\{g_1, \ldots, g_n\}$ is a basis, then every $g \in G$ has a unique representation:

$$g = m_1 g_1 + \ldots + m_n g_n \qquad (m_i \in \mathbf{Z})$$

Because an alternative expression

$$g = k_1 g_1 + \ldots + k_n g_n \qquad (k_i \in \mathbf{Z})$$

implies

$$(m_1 - k_1)g_1 + \ldots + (m_n - k_n)g_n = 0$$

and linear independence implies $m_i = k_i$ $(1 \leqslant i \leqslant n)$.

If $\mathbf{Z}^n$ denotes the direct product of n copies of the additive group of integers, it follows that a group with a basis of n elements is isomorphic to $\mathbf{Z}^n$.

To show that two different bases of G have the same number of elements, let $2G$ be the subgroup of G consisting of all elements of the form $g + g$ ($g \in G$). If G has a basis of n elements, then $G/2G$ is a group of order 2^n. Since the definition of $2G$ does not depend on any particular basis, every basis must have the same number of elements.

An abelian group with a basis of n elements is called a *free abelian group* of *rank n*. If G is free abelian of rank n and $\{x_1, \ldots, x_n\}$, $\{y_1, \ldots, y_n\}$ are both bases, then there exist integers a_{ij}, b_{ij} such that

$$y_i = \sum_j a_{ij}x_j, \qquad x_i = \sum_j b_{ij}y_j.$$

If we consider the matrices

$$A = (a_{ij}), \qquad B = (b_{ij})$$

it follows that $AB = I_n$, the identity matrix. Hence

$$\det(A)\det(B) = 1$$

and since $\det(A)$ and $\det(B)$ are integers, we must have

$$\det(A) = \det(B) = \pm 1.$$

A square matrix over $\mathbf{Z}$ with determinant ± 1 is said to be *unimodular*. We have:

Lemma 1.11. *Let G be a free abelian group of rank n with basis $\{x_1, \ldots, x_n\}$. Suppose (a_{ij}) is an $n \times n$ matrix with integer entries. Then the elements*

$$y_i = \sum_j a_{ij}x_j$$

form a basis of G if and only if (a_{ij}) is unimodular.

Proof. The 'only if' part has already been dealt with. Now suppose $A = (a_{ij})$ is unimodular. Since $\det(A) \neq 0$ it follows that the y_j are linearly independent. We have

$$A^{-1} = (\det(A))^{-1}\tilde{A}$$

where $\tilde{A}$ is the adjoint matrix and has integer entries. Hence $A^{-1} = \pm\tilde{A}$ has integer entries. Putting $B = A^{-1} = (b_{ij})$ we obtain $x_i = \Sigma_j b_{ij}y_j$, demonstrating that the y_j generate A. Thus they form a basis. □

The central result in the theory of finitely generated free abelian groups concerns the structure of subgroups:

Theorem 1.12. *Every subgroup H of a free abelian group G of rank n is free of rank $s \leqslant n$. Moreover there exists a basis $u_1, \ldots, u_n$ for G and positive integers $\alpha_1, \ldots, \alpha_s$ such that $\alpha_1 u_1, \ldots, \alpha_s u_s$ is a basis for H.*

Proof. We use induction on the rank n of G. For $n = 1$, G is infinite cyclic and the result is a consequence of the subgroup structure of the cyclic group. If G is rank n, pick any basis $w_1, \ldots, w_n$ of G. Every $h \in H$ is of the form

$$h = h_1 w_1 + \ldots + h_n w_n.$$

Either $H = \{0\}$, in which case the theorem is trivial, or there exist non-zero coefficients h_i for some $h \in H$. From all such coefficients, let $\lambda(w_1, \ldots, w_n)$ be the least positive integer occurring. Now choose the basis $w_1, \ldots, w_n$ to make $\lambda(w_1, \ldots, w_n)$ minimal. Let α_1 be this minimal value, and number the w_i in such a way that

$$v_1 = \alpha_1 w_1 + \beta_2 w_2 + \ldots + \beta_n w_n$$

is an element of H in which α_1 occurs as a coefficient.
Let

$$\beta_i = \alpha_1 q_i + r_i \qquad (2 \leqslant i \leqslant n)$$

where $0 \leqslant r_i < \alpha_1$, so that r_i is the remainder on dividing β_i by α_1. Define

$$u_1 = w_1 + q_2 w_2 + \ldots + q_n w_n.$$

Then it is easy to verify that $u_1, w_2, \ldots, w_n$ is another basis for G. (The appropriate matrix is clearly unimodular.) With respect to the new basis,

$$v_1 = \alpha_1 u_1 + r_2 w_2 + \ldots + r_n w_n.$$

By the minimality of $\alpha_1 = \lambda(w_1, \ldots, w_n)$ for *all* bases we have

$$r_2 = \ldots = r_n = 0.$$

Hence

$$v_1 = \alpha_1 u_1.$$

With respect to the new basis, let

$$H' = \{m_1 u_1 + m_2 w_2 + \ldots + m_n w_n \mid m_1 = 0\}.$$

Clearly $H' \cap V_1 = \{0\}$, where V_1 is the subgroup generated by v_1. We claim that $H = H' + V_1$. For if $h \in H$ then

$$h = \gamma_1 u_1 + \gamma_2 w_2 + \ldots + \gamma_n w_n$$

and putting

$$\gamma_1 = \alpha_1 q + r_1 \qquad (0 \leqslant r_1 < \alpha_1)$$

it follows that H contains

$$h - qv_1 = r_1 u_1 + \gamma_2 w_2 + \ldots + \gamma_n w_n$$

and the minimality of α_1 once more implies that $r_1 = 0$. Hence $h - qv_1 \in H'$. It follows that H is isomorphic to $H' \times V_1$ and H' is a subgroup of the group G' which is free abelian of rank $n - 1$ with generators $w_2, \ldots, w_n$. By induction, H' is free of rank $\leqslant n - 1$, and there exist bases $u_2, \ldots, u_n$ of G' and $v_2, \ldots, v_s$ of H' such that $v_i = \alpha_i u_i$ for positive integers α_i. The result follows. □

From the above two results we can deduce a useful theorem about orders of quotient groups. In its statement we use $|X|$ to denote the cardinality of the set X, and $|x|$ to denote the absolute value of the real number x. No confusion need arise.

Theorem 1.13. *Let G be a free abelian group of rank r, and H a subgroup of G. Then G/H is finite if and only if the ranks* *of G and H are equal. If this is the case, and if G and H have* **Z**-*bases* $x_1, \ldots, x_r$ *and* $y_1, \ldots, y_r$ *with* $y_i = \Sigma_j a_{ij}x_j$, *then*

$$|G/H| = |\det(a_{ij})|.$$

Proof. Let H have rank s, and use Theorem 1.12 to choose **Z**-bases $u_1, \ldots, u_r$ of G and $v_1, \ldots, v_s$ of H with $v_i = \alpha_i u_i$ for $1 \leqslant i \leqslant s$. Clearly G/H is the direct product of finite cyclic groups of orders $\alpha_1, \ldots, \alpha_s$ and $r - s$ infinite cyclic groups. Hence $|G/H|$ is finite if and only if $r = s$, and in that case

$$|G/H| = \alpha_1 \ldots \alpha_r.$$

Now we have

$$u_i = \sum_j b_{ij}x_j$$

$$v_i = \sum_j c_{ij}u_j$$

$$y_i = \sum_j d_{ij}v_j$$

where the matrices $(b_{ij}) = B$ and $(d_{ij}) = D$ are unimodular by Lemma 1.11, and

$$C = (c_{ij}) = \begin{bmatrix} \alpha_1 & & & & \\ & \alpha_2 & & 0 & \\ & & \cdot & & \\ & 0 & & \cdot & \\ & & & & \alpha_r \end{bmatrix}.$$

Clearly if $A = (a_{ij})$ we have $A = BCD$, and hence

$$\det(A) = \det(B)\det(C)\det(D).$$

So

$$|\det(A)| = |\pm 1|\,|\det(C)|\,|\pm 1| = |\alpha_1 \ldots \alpha_r| = |G/H|$$

as claimed. □

For example, if G has rank 3 and **Z**-basis x, y, z; and if H has **Z**-basis

$$\begin{aligned} &3x + y - 2z, \\ &4x - 5y + z, \\ &x \quad\quad + 7z, \end{aligned}$$

then $|G/H|$ is the absolute value of

$$\begin{vmatrix} 3 & 1 & -2 \\ 4 & -5 & 1 \\ 1 & 0 & 7 \end{vmatrix},$$

namely 142.

Suppose now that G is a finitely generated group, generated by $w_1, \ldots, w_n$ where the latter need not be indepenent. Then we can define a map $f: \mathbf{Z}^n \to G$ by:

$$f(m_1, \ldots, m_n) = m_1 w_1 + \ldots + m_n w_n.$$

This is surjective, so G is isomorphic to $\mathbf{Z}^n/H$ where H is the kernel of f. We can use Theorem 1.12 to choose a new basis $u_1, \ldots, u_n$ of $\mathbf{Z}^n$ in such a way that $\alpha_1 u_1, \ldots, \alpha_s u_s$ is a basis for H. Let A be the subgroup of $\mathbf{Z}^n$ generated by $u_1, \ldots, u_s$ and B be the subgroup generated by $u_{s+1}, \ldots, u_n$, then clearly G is isomorphic to $(A/H) \times B$, and so is the direct product of a finite abelian group A/H and a free group B on $n - s$ generators. Putting $n - s = k$, we have:

Proposition 1.14. *Every finitely generated abelian group with n generators is the direct product of a finite abelian group and a free group on k generators where $k \leqslant n$.* □

If K is any subgroup of a finitely generated abelian group G, then writing $G = F \times B$ where F is finite and B is finitely generated and free, we find $K \cong (F \cap K) \times H$ where $H \subseteq B$. Then $F \cap K$ is finite and (by Theorem 1.12) H is finitely generated and free, so we find K is finitely generated. Hence we have:

Proposition 1.15. *A subgroup of a finitely generated abelian group is finitely generated.* □

Of course the results in this section are not the best possible that can be proved in finitely generated abelian group theory. Refinements may be found in [HH] Chapter 10 (deduced from more general theorems about modules) or proved directly in Ledermann [31]. The results that we have established are ample for our needs in this text, and we will delay no longer in making a start on the substance of algebraic number theory.

Exercises

1.1 Show that Theorem 1.1 becomes false if the word 'finite' is omitted from the hypotheses.

1.2 Which of the following polynomials over $\mathbf{Z}$ are irreducible?
(a) $x^2 + 3$
(b) $x^2 - 169$
(c) $x^3 + x^2 + x + 1$
(d) $x^3 + 2x^2 + 3x + 4$.

1.3 Write down some polynomials over $\mathbf{Z}$ and factorize them into irreducibles.

1.4 Does Theorem 1.2 remain true over a field of characteristic $p > 0$?

1.5 Find the minimum polynomial over $\mathbf{Q}$ of
(i) $(1 + i)/\sqrt{2}$ (ii) $i + \sqrt{2}$ (iii) $e^{2\pi i/3} + 2$.

1.6 Find the degrees of the following field extensions:
(a) $\mathbf{Q}(\sqrt{7}):\mathbf{Q}$
(b) $\mathbf{C}(\sqrt{7}):\mathbf{C}$
(c) $\mathbf{Q}(\sqrt{5},\sqrt{7},\sqrt{35}):\mathbf{Q}$
(d) $\mathbf{R}(\theta):\mathbf{R}$ where $\theta^3 - 7\theta + 6 = 0$ and $\theta \notin \mathbf{R}$.
(e) $\mathbf{Q}(\pi):\mathbf{Q}$.

1.7 Let K be the field generated by the elements $e^{2\pi i/n}$

($n = 1, 2, \ldots$). Show that K is an algebraic extension of $\mathbf{Q}$, but that $[K:\mathbf{Q}]$ is not finite. (It may help to show that the minimum polynomial of $e^{2\pi i/p}$ for p prime is $t^{p-1} + t^{p-2} + \ldots + 1$.)

1.8 Express the following polynomials in terms of elementary symmetric polynomials, where this is possible.
(a) $t_1^2 + t_2^2 + t_3^2$ $(n = 3)$
(b) $t_1^3 + t_2^3$ $(n = 2)$
(c) $t_1 t_2^2 + t_2 t_3^2 + t_3 t_1^2$ $(n = 3)$
(d) $t_1 + t_2^2 + t_3^3$ $(n = 3)$.

1.9 A polynomial belonging to $\mathbf{Z}[t_1, \ldots, t_n]$ is said to be *antisymmetric* if it is invariant under even permutations of the variables, but changes sign under odd permutations. Let

$$\Delta = \prod_{i<j} (t_i - t_j).$$

Show that Δ is antisymmetric. If f is any antisymmetric polynomial, prove that f is expressible as a polynomial in the elementary symmetric polynomials, together with Δ. (*Hint*: consider f/Δ.)

1.10 Find the orders of the groups G/H where G is free abelian with $\mathbf{Z}$-basis, x, y, z and H is generated by:
(a) $2x, 3y, 7z$
(b) $x + 3y - 5z, 2x - 4y, 7x + 2y - 9z$
(c) x
(d) $41x + 32y - 999z, 16y + 3z, 2y + 111z$
(e) $41x + 32y - 999z$.

1.11 Let K be a field. Show that M is a K-module if and only if it is a vector space over K. Show that the submodules of are precisely the vector subspaces. Do these statements remain true if we do not use convention (d) for modules?

1.12 Let **Z** be a **Z**-module with the obvious action. Find all the submodules.

1.13 Let R be a ring, and let M be a finitely generated R-module. Is it true that M necessarily has only finitely many distinct R-submodules? If not, is there an extra condition on R which will lead to this conclusion?

1.14 An abelian group G is said to be *torsion-free* if $g \in G$, $g \neq 0$ and $kg = 0$ for $k \in \mathbf{Z}$ implies $k = 0$. Prove that a finitely generated torsion-free abelian group is a finitely generated free group.

1.15 By examining the proof of Theorem 1.12 carefully, or by other means, prove that if H is a subgroup of a free group G of rank n then there exists a basis $u_1, \ldots, u_n$ for G and a basis $v_1, \ldots, v_s$ for H where $s \leqslant n$ and $v_i = \alpha_i u_i$ $(1 \leqslant i \leqslant s)$ where the α_i are positive integers and α_i divides α_{i+1} $(1 \leqslant i \leqslant s-1)$.

CHAPTER TWO

Algebraic numbers

The factorization of a number depends very much on the ring of which it is considered to be an element. For example, 2 is irreducible in $\mathbf{Z}$, a unit in $\mathbf{Q}$, yet in $\mathbf{Z}[\sqrt{2}]$ we can write it as

$$2 = \sqrt{2}.\sqrt{2}.$$

In $\mathbf{Z}[\sqrt{2}]$ we find $\sqrt{2}$ is not a unit, so here 2 is reducible.

For a suitable generalization of factorization of integers we work in appropriate subrings of $\mathbf{C}$. In this chapter the main characters of the drama make their appearance. They are: an algebraic number field and its ring of algebraic integers. It is in such a ring of algebraic integers that we shall seek a theory of factorization in Chapter 4.

First we define algebraic numbers and algebraic number fields, and prove that each such field is of the form $\mathbf{Q}[\theta]$ for a single algebraic number θ. We introduce the conjugates of an algebraic number and the discriminant of a basis for $\mathbf{Q}[\theta]$ over $\mathbf{Q}$, using the conjugates of θ to show that the discriminant is always a non-zero rational number. Algebraic integers are defined and shown to form a ring. The ring of algebraic integers in a number field is shown to have an integral basis whose discriminant is an integer. This integer is independent of the choice of integral basis and is called the discriminant of the number field.

Finally, we introduce the norm and trace of an algebraic number which prove to be ordinary integers when the algebraic number is an algebraic integer. Using the norm and trace in later chapters we shall be able to translate statements about algebraic integers into statements about ordinary integers which are easier to handle.

2.1 Algebraic numbers

A complex number α will be called *algebraic* if it is algebraic over $\mathbf{Q}$, that is, it satisfies a non-zero polynomial equation with coefficients in $\mathbf{Q}$. Equivalently (clearing out denominators) we may assume the coefficients to be in $\mathbf{Z}$. We let $\mathbf{A}$ denote the set of algebraic numbers. In fact $\mathbf{A}$ is a field, by virtue of:

Theorem 2.1. *The set* $\mathbf{A}$ *of algebraic numbers is a subfield of the complex field* $\mathbf{C}$.

Proof. We use Theorem 1.8, which in this case says that α is algebraic if and only if $[\mathbf{Q}(\alpha):\mathbf{Q}]$ is finite. Suppose that α, β are algebraic. Then

$$[\mathbf{Q}(\alpha, \beta):\mathbf{Q}] = [\mathbf{Q}(\alpha, \beta):\mathbf{Q}(\alpha)]\,[\mathbf{Q}(\alpha):\mathbf{Q}].$$

Now since β is algebraic over $\mathbf{Q}$ it is certainly algebraic over $\mathbf{Q}(\alpha)$, so the first factor on the right is finite; and the second factor is also finite. Hence $[\mathbf{Q}(\alpha, \beta):\mathbf{Q}]$ is finite. But each of $\alpha+\beta$, $\alpha-\beta$, $\alpha\beta$, and (for $\beta \neq 0$) α/β belongs to $\mathbf{Q}(\alpha, \beta)$. So all of these are in $\mathbf{A}$, and the theorem is proved. □

The whole field $\mathbf{A}$ is not as interesting, for us, as certain of its subfields. We define a *number field* to be a subfield K of $\mathbf{C}$ such that $[K:\mathbf{Q}]$ is finite. This implies that every element of K is algebraic, and hence $K \subseteq \mathbf{A}$. The trouble with $\mathbf{A}$ is that $[\mathbf{A}:\mathbf{Q}]$ is not finite (see Exercise 1.7 above, or [GT], Exercise 4.8, p. 55). If K is a number field then $K = \mathbf{Q}(\alpha_1, \ldots, \alpha_n)$ for finitely many algebraic numbers $\alpha_1, \ldots, \alpha_n$ (for instance, a basis for K as vector space over $\mathbf{Q}$). We can strengthen this observation considerably:

Theorem 2.2. *If K is a number field then $K = \mathbb{Q}(\theta)$ for some algebraic number θ.*

Proof. Arguing by induction, it is sufficient to prove that if $K = K_1(\alpha, \beta)$ then $K = K_1(\theta)$ for some θ, (where K_1 is a subfield of K). Let p and q respectively be the minimum polynomials of α, β over K_1, and suppose that over $\mathbf{C}$ these factorize as

$$p(t) = (t - \alpha_1) \ldots (t - \alpha_n),$$
$$q(t) = (t - \beta_1) \ldots (t - \beta_m),$$

where we choose the numbering so that $\alpha_1 = \alpha$, $\beta_1 = \beta$. By Corollary 1.3 the α_i are distinct, as are the β_j. Hence for each i and each $k \neq 1$ there is at most one element $x \in K_1$ such that

$$\alpha_i + x\beta_k = \alpha_1 + x\beta_1.$$

Since there are only finitely many such equations, we may choose $c \neq 0$ in K_1, not equal to any of these x's, and then

$$\alpha_i + c\beta_k \neq \alpha_1 + c\beta_1$$

for $1 \leqslant i \leqslant n$, $2 \leqslant k \leqslant m$. Define

$$\theta = \alpha + c\beta.$$

We shall prove that $K_1(\theta) = K_1(\alpha, \beta)$. Obviously $K_1(\theta) \subseteq K_1(\alpha, \beta)$, and it suffices to prove that $\beta \in K_1(\theta)$ since $\alpha = \theta - c\beta$.

Now
$$p(\theta - c\beta) = p(\alpha) = 0.$$

We define the polynomial

$$r(t) = p(\theta - ct) \in K_1(\theta)[t]$$

and then β is a zero of both $q(t)$ and $r(t)$ as polynomials over $K_1(\theta)$. Now these polynomials have only one common zero, for if $q(\xi) = r(\xi) = 0$ then ξ is one of $\beta_1, \ldots, \beta_m$ and also $\theta - c\xi$ is one of $\alpha_1, \ldots, \alpha_n$. Our choice of c forces $\xi = \beta$. Let $h(t)$ be the minimum polynomial of β over $K_1(\theta)$. Then

$h(t) \mid q(t)$ and $h(t) \mid r(t)$. Since q and r have just one common zero in $\mathbf{C}$ we must have $\partial h = 1$, so that

$$h(t) = t + \mu$$

for $\mu \in K_1(\theta)$. Now $0 = h(\beta) = \beta + \mu$ so that $\beta = -\mu \in K_1(\theta)$ as required. □

Example. $\mathbf{Q}(\sqrt{2}, \sqrt[3]{5})$.

We have

$$\alpha_1 = \sqrt{2}, \alpha_2 = -\sqrt{2},$$
$$\beta_1 = \sqrt[3]{5}, \beta_2 = \omega\sqrt[3]{5}, \beta_3 = \omega^2\sqrt[3]{5}$$

where

$$\omega = \tfrac{1}{2}(-1 + \sqrt{-3})$$

is a complex cube root of 1. The number $c = 1$ satisfies

$$\alpha_i + c\beta_k \neq \alpha + c\beta$$

for $i = 1, 2, k = 2, 3$; since the number on the left is not real in any of the four cases, whereas that on the right is. Hence $\mathbf{Q}(\sqrt{2}, \sqrt[3]{5}) = \mathbf{Q}(\sqrt{2} + \sqrt[3]{5})$.

The expression of K as $\mathbf{Q}(\theta)$ is, of course, not unique; for $\mathbf{Q}(\theta) = \mathbf{Q}(-\theta) = \mathbf{Q}(\theta + 1) = \ldots$ etc.

2.2 Conjugates and discriminants

If $K = \mathbf{Q}(\theta)$ is a number field there will, in general, be several distinct monomorphisms $\sigma : K \to \mathbf{C}$. For instance, if $K = \mathbf{Q}(i)$ where $i = \sqrt{-1}$ then we have the possibilities

$$\sigma_1(x + iy) = x + iy,$$
$$\sigma_2(x + iy) = x - iy,$$

for $x, y \in \mathbf{Q}$. The full set of such monomorphisms will play a fundamental part in the theory, so we begin with a description.

Theorem 2.3. *Let $K = \mathbf{Q}(\theta)$ be a number field of degree n over $\mathbf{Q}$. Then there are exactly n distinct monomorphisms*

$\sigma_i : K \to \mathbf{C}$ $(i = 1, \ldots, n)$. *The elements* $\sigma_i(\theta) = \theta_i$ *are the distinct zeros in* $\mathbf{C}$ *of the minimum polynomial of* θ *over* $\mathbf{Q}$.

Proof. Let $\theta_1, \ldots, \theta_n$ be the (by Corollary 1.3 distinct) zeros of the minimum polynomial p of θ. Then each θ_i also has minimum polynomial p (it must divide p, and p is irreducible) and so there is a unique field isomorphism $\sigma_i : \mathbf{Q}(\theta) \to \mathbf{Q}(\theta_i)$ such that $\sigma_i(\theta) = \theta_i$. In fact, if $\alpha \in \mathbf{Q}(\theta)$ then $\alpha = r(\theta)$ for a unique $r \in \mathbf{Q}[t]$ with $\partial r < n$; and we must have

$$\sigma_i(\alpha) = r(\theta_i).$$

(See [GT], Theorem 3.8, p. 43.) Conversely if $\sigma : K \to \mathbf{C}$ is a monomorphism then σ is the identity on $\mathbf{Q}$. Then we have

$$0 = \sigma(p(\theta)) = p(\sigma(\theta))$$

so that $\sigma(\theta)$ is one of the θ_i, hence σ is one of the σ_i. □

Keep this notation, and for each $\alpha \in K = \mathbf{Q}(\theta)$ define the *field polynomial* of α over K to be

$$f_\alpha(t) = \prod_{i=1}^{n} (t - \sigma_i(\alpha)).$$

As it stands, this is in $K[t]$. In fact more is true:

Theorem 2.4. *The coefficients of the field polynomial are rational numbers, so that* $f_\alpha(t) \in \mathbf{Q}[t]$.

Proof. We have $\alpha = r(\theta)$ for $r \in K[t]$, $\partial r < n$. Now the field polynomial takes the form

$$f_\alpha(t) = \prod_i (t - r(\theta_i))$$

where the θ_i run through all zeros of the minimum polynomial p of θ, whose coefficients are in $\mathbf{Q}$. It is easy to see that the coefficients of $f_\alpha(t)$ are of the form

$$h(\theta_1, \ldots, \theta_n)$$

where $h(t_1, \ldots, t_n)$ is a symmetric polynomial in $\mathbf{Q}[t_1, \ldots, t_n]$. By Corollary 1.10 the result follows. □

The elements $\sigma_i(\alpha)$, for $i = 1, \ldots, n$, are called the *K*-conjugates of α. Although the θ_i are distinct (and are the *K*-conjugates of θ) it is not always the case that the *K*-conjugates of α are distinct: for instance $\sigma_i(1) = 1$ for all i. The precise situation is given by:

Theorem 2.5. *With the above notation,*

(a) *The field polynomial f_α is a power of the minimum polynomial p_α,*

(b) *The K-conjugates of α are the zeros of p_α in $\mathbf{C}$, each repeated n/m times where $m = \partial p_\alpha$ is a divisor of n,*

(c) *The element $\alpha \in \mathbf{Q}$ if and only if all of its K-conjugates are equal,*

(d) *$\mathbf{Q}(\alpha) = \mathbf{Q}(\theta)$ if and only if all K-conjugates of α are distinct.*

Proof. The main point is (a). Now $q = p_\alpha$ is irreducible, and α is a zero of $f = f_\alpha$, so that $f = q^s h$ where q and h are coprime and both are monic. (This follows from factorizing f into irreducibles.) We claim that h is constant. If not, some $\alpha_i = \sigma_i(\alpha) = r(\theta_i)$ is a zero of h, where $\alpha = r(\theta)$. Hence if $g(t) = h(r(t))$ then $g(\theta_i) = 0$. Let p be the minimum polynomial of θ over $\mathbf{Q}$, and hence also of each θ_i. Then $p \mid g$, so that $g(\theta_j) = 0$ for all j, and in particular $g(\theta) = 0$. Therefore, $h(\alpha) = h(r(\theta)) = g(\theta) = 0$ and so q divides h, a contradiction. Hence h is constant and monic, so $h = 1$ and $f = q^s$.

(b) is an immediate consequence of (a) on referring to the definition of the field polynomial.

To prove (c), it is clear that $\alpha \in \mathbf{Q}$ implies $\sigma_i(\alpha) \in \mathbf{Q}$. Conversely if all $\sigma_i(\alpha)$ are equal then, since the zeros of $q = p_\alpha$ are distinct and $f_\alpha = q^s$, then $\partial q = 1$ and so $\alpha \in \mathbf{Q}$.

Finally for (d): if all $\sigma_i(\alpha)$ are distinct then $\partial p_\alpha = n$, and hence $[\mathbf{Q}(\alpha) : \mathbf{Q}] = n = [\mathbf{Q}(\theta) : \mathbf{Q}]$. This implies that $\mathbf{Q}(\alpha) = \mathbf{Q}(\theta)$. Conversely if $\mathbf{Q}(\alpha) = \mathbf{Q}(\theta)$ then $\partial p_\alpha = n$ and so the $\sigma_i(\alpha)$ are distinct. □

Warning. Note that the K-conjugates of α need not be elements of K. Even the θ_i need not be elements of K. For example, let θ be the real cube root of 2. Then $\mathbb{Q}(\theta)$ is a subfield of $\mathbb{R}$. The K-conjugates of θ, however, are $\theta, \omega\theta, \omega^2\theta$, where $\omega = \frac{1}{2}(-1+\sqrt{-3})$. The last two of these are non-real, hence do not lie in $\mathbb{Q}(\theta)$.

Still with $K = \mathbb{Q}(\theta)$ of degree n, let $\{\alpha_1, \ldots, \alpha_n\}$ be a basis of K (as vector space over $\mathbb{Q}$). We define the *discriminant* of this basis to be

$$\Delta[\alpha_1, \ldots, \alpha_n] = \{\det [\sigma_i(\alpha_j)]\}^2. \tag{1}$$

If we pick another basis $\{\beta_1, \ldots, \beta_n\}$ then

$$\beta_k = \sum_{i=1}^{n} c_{ik}\alpha_i \qquad (c_{ik} \in \mathbb{Q})$$

for $k = 1, \ldots, n$, and

$$\det (c_{ik}) \neq 0.$$

The product formula for determinants, and the fact that the σ_i are monomorphisms (and hence the identity on $\mathbb{Q}$) shows that

$$\Delta[\beta_1, \ldots, \beta_n] = [\det (c_{ik})]^2 \Delta[\alpha_1, \ldots, \alpha_n].$$

Theorem 2.6. *The discriminant of any basis for $K = \mathbb{Q}(\theta)$ is rational and non-zero. If all K-conjugates of θ are real then the discriminant of any basis is positive.*

Proof. First we pick a basis with which we can compute: the obvious one is $\{1, \theta, \ldots, \theta^{n-1}\}$. If the conjugates of θ are $\theta_1, \ldots, \theta_n$ then

$$\Delta[1, \theta, \ldots, \theta^{n-1}] = (\det \theta_i^j)^2.$$

A determinant of the form $D = \det (t_i^j)$ is called a *Vandermonde* determinant, and has value

$$D = \prod_{1 \leqslant i < j \leqslant n} (t_i - t_j). \tag{2}$$

To see this, think of everything as lying inside $\mathbf{Q}[t_1, \ldots, t_n]$. Then for $t_i = t_j$ the determinant has two equal rows, so vanishes. Hence D is divisible by each $(t_i - t_j)$. To avoid repeating such a factor twice we take $i < j$. Then comparison of degrees easily shows that D has no other non-constant factors; comparing coefficients of $t_1 t_2^2 \ldots t_n^n$ gives (2).

Hence

$$\Delta = \Delta[1, \theta, \ldots, \theta^{n-1}] = [\prod (\theta_i - \theta_j)]^2.$$

Now D is antisymmetric in the t_i, so that D^2 is symmetric. Hence by the usual argument on symmetric polynomials (Corollary 1.10), Δ is rational. Since the θ_i are distinct, $\Delta \neq 0$.

Now let $\{\beta_1, \ldots, \beta_n\}$ be any basis. Then

$$\Delta[\beta_1, \ldots, \beta_n] = (\det c_{ik})^2 \Delta$$

for certain rational numbers c_{ik}, and $\det(c_{ik}) \neq 0$ so that

$$\Delta[\beta_1, \ldots, \beta_n] \neq 0,$$

and is rational. Clearly if all θ_i are real then Δ is a positive real number, hence so is $\Delta[\beta_1, \ldots, \beta_n]$. □

With the above notation, Δ vanishes if and only if some θ_i is equal to another θ_j. Hence the non-vanishing of Δ allows us to 'discriminate' the θ_i, which motivates calling Δ the discriminant.

2.3 Algebraic integers

A complex number θ is an *algebraic integer* if there is a *monic* polynomial $p(t)$ with integer coefficients such that $p(\theta) = 0$. In other words,

$$\theta^n + a_{n-1}\theta^{n-1} + \ldots + a_0 = 0$$

where $a_i \in \mathbf{Z}$ for all i.

For example, $\theta = \sqrt{-2}$ is an algebraic integer, since $\theta^2 + 2 = 0$; $\tau = \frac{1}{2}(1 + \sqrt{5})$ is an algebraic integer, since

$\tau^2 - \tau - 1 = 0$. But $\phi = 22/7$ is not. It satisfies equations like $7\phi - 22 = 0$, but this is not monic; or like $\phi - 22/7 = 0$, whose coefficients are not integers; but it can be shown without difficulty that ϕ does not satisfy any monic polynomial equation with integer coefficients.

We write **B** for the set of algebraic integers. One of our aims is to prove that **B** is a subring of **A**. We prepare for this by proving:

Lemma 2.7. *A complex number θ is an algebraic integer if and only if the additive group generated by all powers* $1, \theta, \theta^2, \ldots$ *is finitely generated.*

Proof. If θ is an algebraic integer, then for some n we have

$$\theta^n + a_{n-1}\theta^{n-1} + \ldots + a_0 = 0 \tag{3}$$

where the $a_i \in \mathbf{Z}$. We claim that every power of θ lies in the additive group generated by $1, \theta, \ldots, \theta^{n-1}$. Call this group Γ. Then (3) shows that $\theta^n \in \Gamma$. Inductively, if $m \geqslant n$ and $\theta^m \in \Gamma$ then

$$\begin{aligned} \theta^{m+1} &= \theta^{m+1-n}\theta^n \\ &= \theta^{m+1-n}(-a_{n-1}\theta^{n-1} - \ldots - a_0) \in \Gamma. \end{aligned}$$

This proves that every power of θ lies in Γ, which gives one implication.

For the converse, suppose that every power of θ lies in a finitely generated additive group G. The subgroup Γ of G generated by the powers $1, \theta, \theta^2, \ldots, \theta^n$ must also be finitely generated (Proposition 1.15), so we will suppose that Γ has generators $v_1, \ldots, v_n$. Each v_i is a polynomial in θ with integer coefficients, so θv_i is also such a polynomial. Hence there exist integers b_{ij} such that

$$\theta v_i = \sum_{j=1}^{n} b_{ij} v_j.$$

This leads to a system of homogeneous equations for the v_i of the form

$$\begin{aligned}(b_{11}-\theta)v_1+b_{12}v_2+\ldots+b_{1n}v_n&=0\\ b_{21}v_1+(b_{22}-\theta)v_2+\ldots+b_{2n}v_n&=0\\ \ldots\ldots\ldots\ldots\ldots\ldots\ldots\ldots&\\ b_{n1}v_1+b_{n2}v_2+\ldots+(b_{nn}-\theta)v_n&=0.\end{aligned}$$

Since there exists a solution $v_1, \ldots, v_n \in \mathbf{C}$, not all zero, it follows that the determinant

$$\begin{vmatrix} b_{11}-\theta & b_{12} & \ldots & b_{1n} \\ b_{21} & b_{22}-\theta & \ldots & b_{2n} \\ \ldots & \ldots & \ldots & \ldots \\ b_{n1} & b_{n2} & \ldots & b_{nn}-\theta \end{vmatrix}$$

is zero. Expanding this, we see that θ satisfies a monic polynomial equation with integer coefficients. □

Theorem 2.8. *The algebraic integers form a subring of the field of algebraic numbers.*

Proof. Let $\theta, \phi \in \mathbf{B}$. We have to show that $\phi + \theta$ and $\theta\phi \in \mathbf{B}$. By Lemma 2.7 all powers of θ lie in a finitely generated additive subgroup Γ_θ of $\mathbf{C}$, and all powers of ϕ lie in a finitely generated additive subgroup Γ_ϕ. But now all powers of $\theta + \phi$ and of $\theta\phi$ are integer linear combinations of elements $\theta^i\phi^j$ which lie in $\Gamma_\theta\Gamma_\phi \subseteq \mathbf{C}$. But if Γ_θ has generators $v_1, \ldots, v_n$ and Γ_ϕ has generators $w_1, \ldots, w_m$, then $\Gamma_\theta\Gamma_\phi$ is the additive group generated by all v_iw_j for $1 \leqslant i \leqslant n$, $1 \leqslant j \leqslant m$. Hence all powers of $\theta + \phi$ and of $\theta\phi$ lie in a finitely generated additive subgroup of $\mathbf{C}$, so by Lemma 2.7 $\theta + \phi$ and $\theta\phi$ are algebraic integers. Hence $\mathbf{B}$ is a subring of $\mathbf{A}$. □

A simple extension of this technique allows us to prove the following useful theorem.

Theorem 2.9. *Let θ be a complex number satisfying a monic polynomial equation whose coefficients are algebraic integers. Then θ is an algebraic integer.*

Proof. Suppose that

$$\theta^n + \psi_{n-1}\theta^{n-1} + \ldots + \psi_0 = 0$$

where $\psi_0, \ldots, \psi_{n-1} \in \mathbf{B}$. Then these generate a subring Ψ of $\mathbf{B}$. The argument of Lemma 2.7 shows that all powers of θ lie inside a finitely generated Ψ-submodule M of $\mathbf{C}$, spanned by $1, \theta, \ldots, \theta^{n-1}$. By Theorem 2.8, each ψ_i and all its powers lie inside a finitely generated additive group Γ_i with generators γ_{ij} $(1 \leqslant j \leqslant n_i)$. It follows that M lies inside the additive group generated by all elements

$$\gamma_{1j_1}\gamma_{2j_2} \cdots \gamma_{n-1\,j_{n-1}}\theta^k$$

$(1 \leqslant j_i \leqslant n_i, 0 \leqslant i \leqslant n-1, 0 \leqslant k \leqslant n-1)$, which is a finite set. So M is finitely generated as an additive group, and the theorem follows. □

Theorems 2.8 and 2.9 allow us to construct many new algebraic integers out of known ones. For instance, $\sqrt{2}$ and $\sqrt{3}$ are clearly algebraic integers. Then Theorem 2.8 says that numbers such as $\sqrt{2}+\sqrt{3}$, $7\sqrt{2}-41\sqrt{3}$, $(\sqrt{2})^5(1+\sqrt{3})^2$ are also algebraic integers. And Theorem 2.9 says that zeros of polynomials such as

$$t^{23} - (14 + \sqrt[5]{3})t^9 + (\sqrt[3]{2})t^5 - 19\sqrt{3}$$

are algebraic integers. It would not be easy, particularly in the last instance, to compute explicit polynomials over $\mathbf{Z}$ of which these algebraic integers are zeros; although it can in principle be done by using symmetric polynomials. In fact Theorems 2.8 and 2.9 can be proved this way.

For any number field K we write

$$\mathfrak{O} = K \cap \mathbf{B},$$

and call $\mathfrak{O}$ the *ring of integers* of K. The symbol '$\mathfrak{O}$' is a Gothic capital O (for 'order', the old terminology). In cases where it is not immediately clear which number field is involved, we write more explicitly $\mathfrak{O}_K$. Since K and $\mathbf{B}$ are subrings of $\mathbf{C}$ it follows that $\mathfrak{O}$ is a subring of K. Further $\mathbf{Z} \subseteq \mathbf{Q} \subseteq K$ and $\mathbf{Z} \subseteq \mathbf{B}$ so $\mathbf{Z} \subseteq \mathfrak{O}$.

The following lemma is easy to prove:

Lemma 2.10. *If $\alpha \in K$ then for some non-zero $c \in \mathbf{Z}$ we have $c\alpha \in \mathfrak{O}$.* □

Corollary 2.11. *If K is a number field then $K = \mathbf{Q}(\theta)$ for an algebraic integer θ.*

Proof. We have $K = \mathbf{Q}(\phi)$ for an algebraic number ϕ by Theorem 2.2. By Lemma 2.10, $\theta = c\phi$ is an algebraic integer for some $0 \neq c \in \mathbf{Z}$. Clearly $\mathbf{Q}(\phi) = \mathbf{Q}(\theta)$. □

Warning. For $\theta \in \mathbf{C}$ let us write $\mathbf{Z}[\theta]$ for the set of elements $p(\theta)$, for polynomials $p \in \mathbf{Z}[t]$. If $K = \mathbf{Q}(\theta)$ where θ is an algebraic integer then certainly $\mathfrak{O}$ contains $\mathbf{Z}[\theta]$ since $\mathfrak{O}$ is a ring containing θ. However, $\mathfrak{O}$ need not equal $\mathbf{Z}[\theta]$. For example, $\mathbf{Q}(\sqrt{5})$ is a number field and $\sqrt{5}$ an algebraic integer. But

$$\frac{1+\sqrt{5}}{2}$$

is a zero of $t^2 - t - 1$, hence an algebraic integer; and it lies in $\mathbf{Q}(\sqrt{5})$ so belongs to $\mathfrak{O}$. It does not belong to $\mathbf{Z}[\sqrt{5}]$.

There is a useful criterion, in terms of the minimum polynomial, for a number to be an algebraic integer:

Lemma 2.12. *An algebraic number α is an algebraic integer if and only if its minimum polynomial over $\mathbf{Q}$ has coefficients in $\mathbf{Z}$.*

Proof. Let p be the minimum polynomial of α over $\mathbf{Q}$, and recall that this is monic and irreducible in $\mathbf{Q}[t]$. If $p \in \mathbf{Z}[t]$ then α is an algebraic integer. Conversely, if α is an algebraic integer then $q(\alpha) = 0$ for some monic $q \in \mathbf{Z}[t]$, and $p \mid q$. By Gauss's Lemma 1.4 it follows that $p \in \mathbf{Z}[t]$, because some rational multiple λp lies in $\mathbf{Z}[t]$ and divides q, and the monicity of q and p implies $\lambda = 1$. □

To avoid confusion as to the usage of the word 'integer' we adopt the following convention: a *rational integer* is an element of $\mathbf{Z}$, and a plain *integer* is an algebraic integer. (The aim is to reserve the shorter term for the concept most often encountered.) Any remaining possibility of confusion is eliminated by:

Lemma 2.13. *An algebraic integer is a rational number if and only if it is a rational integer. Equivalently,* $\mathbf{B} \cap \mathbf{Q} = \mathbf{Z}$.

Proof. Clearly $\mathbf{Z} \subseteq \mathbf{B} \cap \mathbf{Q}$. Let $\alpha \in \mathbf{B} \cap \mathbf{Q}$; since $\alpha \in \mathbf{Q}$ its minimum polynomial over $\mathbf{Q}$ is $t - \alpha$. By Lemma 2.12 the coefficients of this are in $\mathbf{Z}$, hence $-\alpha \in \mathbf{Z}$, hence $\alpha \in \mathbf{Z}$. □

2.4 Integral bases

Let K be a number field of degree n (over $\mathbf{Q}$). A *basis* (or $\mathbf{Q}$-*basis* for emphasis) of K is a basis for K as a vector space over $\mathbf{Q}$. By Corollary 2.11 we have $K = \mathbf{Q}(\theta)$ where θ is an algebraic integer, and it follows that the minimum polynomial p of θ has degree n and that $\{1, \theta, \ldots, \theta^{n-1}\}$ is a basis for K.

The ring $\mathfrak{O}$ of integers of K is an abelian group under addition. A $\mathbf{Z}$-basis for $(\mathfrak{O}, +)$ is called an *integral basis* for K (or for $\mathfrak{O}$). Thus $\{\alpha_1, \ldots, \alpha_s\}$ is an integral basis if and only if all $\alpha_i \in \mathfrak{O}$ and every element of $\mathfrak{O}$ is *uniquely* expressible in the form

$$a_1\alpha_1 + \ldots + a_s\alpha_s$$

for rational integers $a_1, \ldots, a_s$. It is obvious from Lemma 2.10 that any integral basis for K is a $\mathbf{Q}$-basis. Hence in particular $s = n$. But we have to verify that integral bases exist. In fact they do, but they are not always what naively we might expect them to be.

For instance we can assert that $K = \mathbf{Q}[\theta]$ $(= \mathbf{Q}(\theta))$ for an algebraic integer θ (Corollary 2.11), so that $\{1, \theta, \ldots, \theta^{n-1}\}$ is a $\mathbf{Q}$-basis for K which consists of integers, but it does not follow that $\{1, \theta, \ldots, \theta^{n-1}\}$ is an integral basis. Some of the elements in $\mathbf{Q}[\theta]$ with rational coefficients may also be integers. As an example, consider $K = \mathbf{Q}(\sqrt{5})$. We have seen that the element $\frac{1}{2} + \frac{1}{2}\sqrt{5}$ satisfies the equation

$$t^2 - t + 1 = 0$$

and so is an integer in $\mathbf{Q}(\sqrt{5})$, but it is not an element of $\mathbf{Z}[\sqrt{5}]$.

Our first problem, therefore, is to show that integral bases exist. That they do is equivalent to the statement that $(\mathfrak{O}, +)$ is a free abelian group of rank n. To prove this we first establish:

Lemma 2.14. *If* $\{\alpha_1, \ldots, \alpha_n\}$ *is a basis of* K *consisting of integers, then the discriminant* $\Delta[\alpha_1, \ldots, \alpha_n]$ *is a rational integer, not equal to zero.*

Proof. We know that $\Delta = \Delta[\alpha_1, \ldots, \alpha_n]$ is rational by Theorem 2.6, and it is an integer since the α_i are. Hence by Lemma 2.13 it is a rational integer. By Theorem 2.6, $\Delta \neq 0$.
□

Theorem 2.15. *Every number field* K *possesses an integral basis, and the additive group of* $\mathfrak{O}$ *is free abelian of rank* n *equal to the degree of* K.

Proof. We have $K = \mathbf{Q}(\theta)$ for θ an integer. Hence there exist bases for K consisting of integers: for example $\{1, \theta, \ldots, \theta^{n-1}\}$. We have already seen that such $\mathbf{Q}$-bases need not be integral bases. However, the discriminant of a $\mathbf{Q}$-basis consisting of integers is always a rational integer (Lemma 2.14), so what we do is to select a basis $\{\omega_1, \ldots, \omega_n\}$ of integers for which

$$|\Delta[\omega_1, \ldots, \omega_n]|$$

is least. We claim that this is in fact an integral basis. If not, there is an integer ω of K such that

$$\omega = a_1\omega_1 + \ldots + a_n\omega_n$$

for $a_i \in \mathbf{Q}$, *not all in* $\mathbf{Z}$. Choose the numbering so that $a_1 \notin \mathbf{Z}$. Then $a_1 = a + r$ where $a \in \mathbf{Z}$ and $0 < r < 1$. Define

$$\psi_1 = \omega - a\omega_1, \qquad \psi_i = \omega_i \qquad (i = 2, \ldots, n).$$

Then $\{\psi_1, \ldots, \psi_n\}$ is a basis consisting of integers. The determinant relevant to the change of basis from the ω's to the ψ's is

$$\begin{vmatrix} a_1 - a & a_2 & a_3 & \ldots & a_n \\ 0 & 1 & 0 & \ldots & 0 \\ 0 & 0 & 1 & \ldots & 0 \\ & \cdots & \cdots & \cdots & \\ 0 & 0 & 0 & \ldots & 1 \end{vmatrix} = r,$$

and so

$$\Delta[\psi_1, \ldots, \psi_n] = r^2 \Delta[\omega_1, \ldots, \omega_n].$$

Since $0 < r < 1$ this contradicts the choice of $\{\omega_1, \ldots, \omega_n\}$ making $|\Delta[\omega_1, \ldots, \omega_n]|$ minimal.

It follows that $\{\omega_1, \ldots, \omega_n\}$ is an integral basis, and so $(\mathfrak{O}, +)$ is free abelian of rank n. □

This raises the question of finding integral bases in cases such as $\mathbf{Q}(\sqrt{5})$ where the $\mathbf{Q}$-basis $\{1, \sqrt{5}\}$ is not an integral basis. We shall consider a more general case in the next chapter, but this particular example is worth a brief discussion here.

An element of $\mathbf{Q}(\sqrt{5})$ is of the form $p + q\sqrt{5}$ for $p, q \in \mathbf{Q}$, and has minimum polynomial

$$(t - p - q\sqrt{5})(t - p + q\sqrt{5}) = t^2 - 2pt + (p^2 - 5q^2).$$

Then $p + q\sqrt{5}$ is an integer if and only if the coefficients $2p, p^2 - 5q^2$ are rational integers. Thus $p = \frac{1}{2}P$ where P is a rational integer. For P even, we have p^2 a rational integer, so $5q^2$ is a rational integer also, implying q is a rational integer. For P odd, a straightforward calculation (performed in the next chapter in greater generality) shows $q = \frac{1}{2}Q$ where Q is also an odd rational integer.

From this it follows that $\mathfrak{O} = \mathbf{Z}[\frac{1}{2} + \frac{1}{2}\sqrt{5}]$ and an integral basis is $\{1, \frac{1}{2} + \frac{1}{2}\sqrt{5}\}$.

We can prove this by another route using the discriminant. The two monomorphisms $\mathbf{Q}(\sqrt{5}) \to \mathbf{C}$ are given by

$$\sigma_1(p + q\sqrt{5}) = p + q\sqrt{5},$$
$$\sigma_2(p + q\sqrt{5}) = p - q\sqrt{5}.$$

Hence the discriminant $\Delta[1, \frac{1}{2} + \frac{1}{2}\sqrt{5}]$ is given by

$$\begin{vmatrix} 1 & \frac{1}{2} + \frac{1}{2}\sqrt{5} \\ 1 & \frac{1}{2} - \frac{1}{2}\sqrt{5} \end{vmatrix}^2 = 5.$$

We define a rational integer to be *squarefree* if it is not divisible by the square of a prime. For example, 5 is squarefree, as are 6, 7, but not 8 or 9. Given a $\mathbf{Q}$-basis of K consisting of integers, we compute the discriminant and then we have:

Theorem 2.16. *Suppose $\alpha_1, \ldots, \alpha_n \in \mathfrak{O}$ form a $\mathbf{Q}$-basis for K. If $\Delta[\alpha_1, \ldots, \alpha_n]$ is squarefree then $\{\alpha_1, \ldots, \alpha_n\}$ is an integral basis.*

Proof. Let $\{\beta_1, \ldots, \beta_n\}$ be an integral basis. Then there exist rational integers c_{ij} such that $\alpha_i = \Sigma c_{ij}\beta_j$, and

$$\Delta[\alpha_1, \ldots, \alpha_n] = (\det c_{ij})^2 \Delta[\beta_1, \ldots, \beta_n].$$

Since the left-hand side is squarefree, we must have $\det c_{ij} = \pm 1$, so that (c_{ij}) is unimodular. Hence by Lemma 1.11 $\{\alpha_1, \ldots, \alpha_n\}$ is a $\mathbf{Z}$-basis for $\mathfrak{O}$, that is, an integral basis for K. □

For example, the $\mathbf{Q}$-basis $\{1, \frac{1}{2} + \frac{1}{2}\sqrt{5}\}$ for $\mathbf{Q}(\sqrt{5})$ consists of integers and has discriminant 5 (calculated above). Since 5 is squarefree, this is an integral basis. The reader should note that there exist integral bases whose discriminants are not squarefree (as we shall see later on), so the converse of Theorem 2.16 is false.

For two integral bases $\{\alpha_1, \ldots, \alpha_n\}$, $\{\beta_1, \ldots, \beta_n\}$ of an algebraic number field K, we have

$$\Delta[\alpha_1, \ldots, \alpha_n] = (\pm 1)^2 \Delta[\beta_1, \ldots, \beta_n] = \Delta[\beta_1, \ldots, \beta_n],$$

because the matrix corresponding to the change of basis is unimodular. Hence the discriminant of an integral basis is independent of which integral basis we choose. This common value is called the *discriminant of* K (or of $\mathfrak{O}$). It is always a non-zero rational integer. Obviously, isomorphic number fields have the same discriminant. The important role played by the discriminant will become apparent as the drama unfolds.

2.5 Norms and traces

These important concepts often allow us to transform a problem about algebraic integers into one about rational integers. As usual, let $K = \mathbf{Q}(\theta)$ be a number field of degree n and let $\sigma_1, \ldots, \sigma_n$ be the monomorphisms $K \to \mathbf{C}$. Now the field polynomial is a power of the minimum polynomial by Theorem 2.5(a), so by Lemma 2.12 and Gauss's Lemma 1.4 it follows that $\alpha \in K$ is an integer if and only if the field polynomial has rational integer coefficients. For any $\alpha \in K$ we define the *norm*

$$N_K(\alpha) = \prod_{i=1}^{n} \sigma_i(\alpha)$$

and *trace*

$$T_K(\alpha) = \sum_{i=1}^{n} \sigma_i(\alpha).$$

Where the field K is clear from the context, we will abbreviate the norm and trace of α to $N(\alpha)$ and $T(\alpha)$ respectively.

Since the field polynomial is

$$f_\alpha(t) = \prod_{i=1}^{n} (t - \sigma_i(\alpha))$$

it follows from the remark above that *if* α *is an integer then the norm and trace of* α *are rational integers.* Since the σ_i are monomorphisms it is clear that

$$N(\alpha\beta) = N(\alpha)N(\beta) \tag{4}$$

and if $\alpha \neq 0$ then $\mathrm{N}(\alpha) \neq 0$. If p, q are rational numbers then

$$\mathrm{T}(p\alpha + q\beta) = p\mathrm{T}(\alpha) + q\mathrm{T}(\beta). \tag{5}$$

For instance, if $K = \mathbf{Q}(\sqrt{7})$ then the integers of K are given by $\mathfrak{O} = \mathbf{Z}[\sqrt{7}]$ (as we shall see in Theorem 3.2). The maps σ_i are given by

$$\sigma_1(p + q\sqrt{7}) = p + q\sqrt{7},$$
$$\sigma_2(p + q\sqrt{7}) = p - q\sqrt{7}.$$

Hence

$$\mathrm{N}(p + q\sqrt{7}) = p^2 - 7q^2,$$
$$\mathrm{T}(p + q\sqrt{7}) = 2p.$$

Since norms are not too hard to compute (they can always be found from symmetric polynomial considerations, often with short-cuts) whereas discriminants involve complicated work with determinants, the following result is sometimes useful:

Proposition 2.17. *Let $K = \mathbf{Q}(\theta)$ be a number field where θ has minimum polynomial p of degree n. The $\mathbf{Q}$-basis $\{1, \theta, \ldots, \theta^{n-1}\}$ has discriminant*

$$\Delta[1, \ldots, \theta^{n-1}] = (-1)^{n(n-1)/2}\mathrm{N}(\mathrm{D}p(\theta))$$

where D*p is the formal derivative of p.*

Proof. From the proof of Theorem 2.6 we obtain

$$\Delta = \Delta[1, \theta, \ldots, \theta^{n-1}] = \prod_{1 \leqslant i < j \leqslant n} (\theta_i - \theta_j)^2$$

where $\theta_1, \ldots, \theta_n$ are the conjugates of θ. Now

$$p(t) = \prod_{i=1}^{n} (t - \theta_i)$$

so that

$$\mathrm{D}p(t) = \sum_{j=1}^{n} \prod_{\substack{i=1 \\ i \neq j}}^{n} (t - \theta_i)$$

and therefore

$$\mathrm{D}p(\theta_j) = \prod_{\substack{i=1\\ i\neq j}}^{n} (\theta_j - \theta_i).$$

Multiplying all these equations for $j = 1, \ldots, n$ we obtain

$$\prod_{j=1}^{n} \mathrm{D}p(\theta_j) = \prod_{\substack{i,j=1\\ i\neq j}}^{n} (\theta_j - \theta_i).$$

The left-hand side is $\mathrm{N}(\mathrm{D}p(\theta))$. On the right, each factor $(\theta_i - \theta_j)$ for $i < j$ appears twice, once as $(\theta_i - \theta_j)$ and once as $(\theta_j - \theta_i)$. The product of these two factors is $-(\theta_i - \theta_j)^2$. On multiplying up, we get Δ multiplied by $(-1)^s$ where s is the number of pairs (i, j) with $1 \leqslant i < j \leqslant n$, which is given by

$$s = \tfrac{1}{2}n(n-1).$$

The result follows. □

We close this chapter by noting the following simple identity linking the discriminant and trace:

Proposition 2.18. *If* $\{\alpha_1, \ldots, \alpha_n\}$ *is any* $\mathbf{Q}$*-basis of* K, *then*

$$\Delta[\alpha_1, \ldots, \alpha_n] = \det(\mathrm{T}(\alpha_i\alpha_j)).$$

Proof. $\mathrm{T}(\alpha_i\alpha_j) = \Sigma_{r=1}^{n} \sigma_r(\alpha_i\alpha_j) = \Sigma_{r=1}^{n} \sigma_r(\alpha_i)\sigma_r(\alpha_j)$. Hence

$$\begin{aligned}\Delta[\alpha_1, \ldots, \alpha_n] &= (\det(\sigma_i(\alpha_j)))^2\\ &= (\det(\sigma_j(\alpha_i)))(\det(\sigma_i(\alpha_j)))\\ &= \det(\Sigma_{r=1}^{n} \sigma_r(\alpha_i)\sigma_r(\alpha_j))\\ &= \det(\mathrm{T}(\alpha_i\alpha_j)).\end{aligned}$$

□

2.6 Rings of integers

We now discuss how to find the ring of integers of a given number field. With the methods available to us, this involves moderately heavy calculation; but by taking advantage of short cuts the technique can be made reasonably efficient. In particular we show in Example 3 below that not every number field has an integral basis of the form $\{1, \theta, \ldots, \theta^{n-1}\}$.

The method is based on the following result:

Theorem 2.19. *Let G be an additive subgroup of $\mathfrak{O}$ of rank equal to the degree of K, with $\mathbf{Z}$-basis $\{\alpha_1, \ldots, \alpha_n\}$. Then $|\mathfrak{O}/G|^2$ divides $\Delta[\alpha_1, \ldots, \alpha_n]$.*

Proof. By Theorem 1.12 there exists a $\mathbf{Z}$-basis for $\mathfrak{O}$ of the form $\{\beta_1, \ldots, \beta_n\}$ such that G has a $\mathbf{Z}$-basis $\{\mu_1\beta_1, \ldots, \mu_n\beta_n\}$ for suitable $\mu_i \in \mathbf{Z}$. Now

$$\Delta[\alpha_1, \ldots, \alpha_n] = \Delta[\mu_1\beta_1, \ldots, \mu_n\beta_n]$$

since by Lemma 1.11 a basis-change has a unimodular matrix; and the right-hand side is equal to

$$\begin{aligned}&(\mu_1 \ldots \mu_n)^2\, \Delta[\beta_1, \ldots, \beta_n]\\ &= (\mu_1 \ldots \mu_n)^2 \Delta\end{aligned}$$

where Δ is the discriminant of K and so lies in $\mathbf{Z}$. But

$$|\mu_1 \ldots \mu_n| = |\mathfrak{O}/G|.$$

Therefore

$$|\mathfrak{O}/G|^2 \text{ divides } \Delta[\alpha_1, \ldots, \alpha_n]. \qquad \square$$

In the above situation we use the notation

$$\Delta_G = \Delta[\alpha_1, \ldots, \alpha_n].$$

We then have a generalization of Theorem 2.16:

Proposition 2.20. *Suppose that $G \neq \mathfrak{O}$. Then there exists an algebraic integer of the form*

$$\frac{1}{p}(\lambda_1\alpha_1 + \ldots + \lambda_n\alpha_n) \qquad (6)$$

where $0 \leqslant \lambda_i \leqslant p-1$, $\lambda_i \in \mathbf{Z}$, and p is a prime such that p^2 divides Δ_G.

Proof. If $G \neq \mathfrak{O}$ then $|\mathfrak{O}/G| > 1$. Therefore (by the structure theory for finite abelian groups) there exists a prime p

dividing $|\mathfrak{O}/G|$ and an element $u \in \mathfrak{O}/G$ such that $g = pu \in G$. By Theorem 2.19, p^2 divides Δ_G. Further,

$$u = \frac{1}{p} g = \frac{1}{p} (\lambda_1 \alpha_1 + \ldots + \lambda_n \alpha_n)$$

since $\{\alpha_i\}$ forms a $\mathbf{Z}$-basis for G.

Note that this really *is* a generalization of Theorem 2.16: if Δ_G is squarefree then no such p exists, so that $G = \mathfrak{O}$.

We may use Proposition 2.20 as the basis of a trial-and-error search for algebraic integers in $\mathfrak{O}$ but not in G, because there are only finitely many possibilities (6). The idea is:

(a) Start with an initial guess G for $\mathfrak{O}$.

(b) Compute Δ_G.

(c) For each prime p whose square divides Δ_G, test all numbers of the form (6) to see which are algebraic integers.

(d) If any new integers arise, enlarge G to a new G' by adding in the new number (and divide Δ_G by p^2 to get $\Delta_{G'}$).

(e) Repeat until no new algebraic integers are found.

Example 1. Find the ring of integers of $\mathbf{Q}(\sqrt[3]{5})$.

Let $\theta \in \mathbb{R}$, $\theta^3 = 5$. The natural first guess is that $\mathfrak{O}$ has $\mathbf{Z}$-basis $\{1, \theta, \theta^2\}$. Let G be the abelian group generated by this set. Let $\omega = e^{2\pi i/3}$ be a cube root of unity. We compute

$$\Delta_G = \begin{vmatrix} 1 & \theta & \theta^2 \\ 1 & \omega\theta & \omega^2\theta^2 \\ 1 & \omega^2\theta & \omega\theta^2 \end{vmatrix}^2$$

$$= \theta^3 \begin{vmatrix} 1 & 1 & 1 \\ 1 & \omega & \omega^2 \\ 1 & \omega^2 & \omega \end{vmatrix}^2$$

$$= 5^2 . (\omega^2 + \omega^2 + \omega^2 - \omega - \omega - \omega)^2$$

$$= 5^2 . 3^2 . (\omega^2 - \omega)^2$$

$$= 3^2 . 5^2 . (-3)$$

$$= -3^3 . 5^2 .$$

By Proposition 2.20 we must consider two possibilities.

(a) Can $\alpha = \frac{1}{3}(\lambda_1 + \lambda_2\theta + \lambda_3\theta^2)$ be an algebraic integer, for $0 \leqslant \lambda_i \leqslant 2$?

(b) Can $\alpha = \frac{1}{5}(\lambda_1 + \lambda_2\theta + \lambda_3\theta^2)$ be an algebraic integer, for $0 \leqslant \lambda_i \leqslant 4$?

Consider case (b), which is harder. First use the trace: we have

$$T(\alpha) = 3\lambda_1/5 \in \mathbf{Z}$$

so that $\lambda_1 \in 5\mathbf{Z}$. Then

$$\alpha' = \tfrac{1}{5}(\lambda_2\theta + \lambda_3\theta^2)$$

is also an algebraic integer.

Now compute the norm of α'. (It is easier to do this for α' than for α because there are fewer terms, which is why we use the trace first.) We have

$$\begin{aligned} N(a\theta + b\theta^2) &= (a\theta + b\theta^2)(a\omega\theta + b\omega^2\theta^2)(a\omega^2\theta + b\omega\theta^2) \\ &= \omega\,.\,\omega^2(a\theta + b\theta^2)(a\theta + \omega b\theta^2)(a\theta + \omega^2 b\theta^2) \\ &= (a\theta)^3 + (b\theta^2)^3 \\ &= 5a^3 + 25b^3. \end{aligned}$$

It follows that for α to be an algebraic integer, we must have $N(\alpha') \in \mathbf{Z}$. But $N(\alpha') = (5\lambda_2^3 + 25\lambda_3^3)/125 = (\lambda_2^3 + 5\lambda_3^3)/25$. One way to finish the calculation is just to try all cases:

λ_2	λ_3	$\lambda_2^3 + 5\lambda_3^3$	Divisible by 25?
0	1	5	NO
0	2	40	NO
0	3	135	NO
0	4	320	NO
1	0	1	NO
1	1	6	NO
1	2	41	NO
1	3	136	NO
1	4	321	NO
2	0	8	NO
2	1	13	NO

2	2	48	NO
2	3	143	NO
2	4	328	NO
3	0	27	NO
3	1	32	NO
3	2	67	NO
3	3	162	NO
3	4	347	NO
4	0	64	NO
4	1	69	NO
4	2	104	NO
4	3	199	NO
4	4	384	NO

Whichever argument we use, we have shown that no new better ideas, brute force can suffice. But here it is not hard to find a better idea. Suppose $\lambda_2^3 + 5\lambda_3^3 \equiv 0 \pmod{25}$. If $\lambda_3 \equiv 0 \pmod 5$, then we must also have $\lambda_2 \equiv 0 \pmod 5$. If not, we have $5 \equiv (-\lambda_2/\lambda_3)^3 \pmod{25}$. Therefore 5 is a *cubic residue* (mod 25), that is, is congruent to a cube. The factor 5 shows that we must have $5 \equiv (5k)^3 \pmod{25}$, but then $5 \equiv 0 \pmod{25}$, an impossibility.

Whichever argument we use, we have shown that no new α' occurs in case (b). The analysis in case (a) is similar, and left as exercise 2.6.

Note that it is *necessary* for $N(\alpha)$ and $T(\alpha)$ to be rational integers, in order for α to be an algebraic integer; but it may not be *sufficient*. If the use of norms and traces produces a candidate for a new algebraic integer, we still have to check that it is one - for example, by finding its minimum polynomial. However, our main use of $N(\alpha)$ and $T(\alpha)$ is to rule out possible candidates, so this step is not always needed.

Example 2.

(a) Find the ring of integers of $\mathbf{Q}(\sqrt[3]{175})$.

(b) Show that it has no $\mathbf{Z}$-basis of the form $\{1, \theta, \theta^2\}$.

(a) Let $t = \sqrt[3]{175} = \sqrt[3]{(5^2.7)}$. Consider also $u = \sqrt[3]{5.7^2)} = \sqrt[3]{245}$. We have

$$\begin{aligned} ut &= 35 \\ u^2 &= 7t \\ t^2 &= 5u. \end{aligned}$$

Let $\mathfrak{O}$ be the ring of integers of $K = \mathbf{Q}(\sqrt[3]{175})$.

We have $u = 35/t \in K$. But $u^3 - 245 = 0$ so $u \in \mathbf{B}$. Therefore $u \in \mathbf{B} \cap K = \mathfrak{O}$.

A good initial guess is that $\mathfrak{O} = G$, where G is the abelian group generated by $\{1, t, u\}$.

To see if this is correct, we compute Δ_G. The monomorphisms $K \to \mathbf{C}$ are $\sigma_1, \sigma_2, \sigma_3$ where $\sigma_1(t) = t$, $\sigma_2(t) = \omega t$, $\sigma_2(t) = \omega^2 t$. Since $tu = 35$ which must be fixed by each σ_i, we have $\sigma_1(u) = u$, $\sigma_2(u) = \omega^2 u$, $\sigma_3(u) = \omega u$. Therefore

$$\Delta_G = \begin{vmatrix} 1 & t & u \\ 1 & \omega t & \omega^2 u \\ 1 & \omega^2 t & \omega u \end{vmatrix}^2$$

which works out as $-3^3 \,.\, 5^2 \,.\, 7^2$.

There are now three primes to try: $p = 3, 5$, or 7.

If $p = 5$ or 7 then, as in Example 1, use of the trace lets us assume that our putative integer is $\frac{1}{p}(at + bu)$ for $a, b, \in \mathbf{Z}$. Now

$$N(at + bu) = 175a^3 + 245b^3$$

and we must see whether this can be congruent to 0 (mod 5^3 or 7^3) for a, b not congruent to zero.

Suppose $175a^3 + 245b^3 \equiv 0 \pmod{125}$, that is, $35a^3 + 49b^3 \equiv 0 \pmod{25}$. Write this as $10a^3 - b^3 \equiv 0 \pmod{25}$. If $a \equiv 0 \pmod 5$ then also $b \equiv 0 \pmod 5$. If not, $10 \equiv (b/a)^3 \pmod{25}$ is a cubic residue; but then $10 \equiv (5k)^3 \pmod{25}$, hence $10 \equiv 0 \pmod{25}$ which is absurd. The case $p = 7$ is dealt with in the same way.

When $p = 3$ the trace is no help, and we must compute the

norm of

$$\tfrac{1}{3}(a + bt + cu)$$

for $a, b, c \in \mathbf{Z}$. The calculation is more complicated, but not too bad since we only have to consider $a, b, c = 0, 1, 2$. No new integers occur.

Therefore $\mathfrak{O} = G$ as we hoped.

Now we have to show that there is no **Z**-basis of the form $\{1, \theta, \theta^2\}$, where $\theta = a + bt + cu$. Note that $\{1, \theta, \theta^2\}$ is a **Z**-basis if and only if $\{1, \theta + 1, (\theta + 1)^2\}$ is a **Z**-basis; so we may without loss of generality assume that $a = 0$. Now

$$\begin{aligned}(bt + cu)^2 &= b^2t^2 + 2bctu + c^2u^2 \\ &= 5b^2u + 70bc + 7c^2t.\end{aligned}$$

Therefore $\{1, bt + cu, (bt + cu)^2\}$ is a **Z**-basis if and only if the matrix

$$\begin{vmatrix} 1 & 0 & 0 \\ 0 & b & c \\ 70bc & 7c^2 & 5b^2 \end{vmatrix}$$

is unimodular; that is,

$$5b^3 - 7c^3 = \pm 1.$$

Consider this modulo 7. Cubes are congruent to 0, 1, or $-1 \pmod 7$, so we have $5(-1, 0, \text{or } 1) \equiv \pm 1 \pmod 7$, a contradiction.

Hence no such **Z**-basis exists.

Example 3. Find the ring of integers of $\mathbf{Q}(\sqrt{2}, i)$.

(Here, our initial guess turns out not to be good enough, so this example illustrates how to continue the analysis when this unfortunate event occurs.)

The obvious guess is $\{1, \sqrt{2}, i, i\sqrt{2}\}$. Let G be the group these generate. We have $\Delta_G = -64$, so $\mathfrak{O}$ may contain elements of the form $\frac{1}{2}g$ (and then possibly $\frac{1}{4}g$ or $\frac{1}{8}g$) for $g \in G$. The norm is

$$N(a + b\sqrt{2} + ci + di\sqrt{2}) = (a^2 - c^2 - 2b^2 + 2d^2)^2 + 4(ac - 2bd)^2 .$$

We must find whether this is divisible by 16 for $a, b, c, d =$ 0 or 1, and not all zero. By trial and error the only case where this occurs is $b = d = 1, a = c = 0$. So

$$\alpha = \tfrac{1}{2}(\theta + \theta i)$$

may be an integer (where $\theta = \sqrt{2}$). In fact

$$\alpha^2 = i$$

so that

$$\alpha^4 + 1 = 0$$

and α *is* an integer.

We therefore revise our initial guess to

$$G' = \{1, \theta, i, \theta i, \tfrac{1}{2}\theta(1 + i)\}.$$

Since $2 . \tfrac{1}{2}\theta(1 + i) = \theta + \theta i$ this has a **Z**-basis

$$\{1, \theta, i, \tfrac{1}{2}\theta(1 + i)\}.$$

Now

$$\Delta_{G'} = -64/2^2 = -16.$$

A recalculation of the usual kind shows that nothing of the form $\tfrac{1}{2}g$ (where we may now assume that the term in $\tfrac{1}{2}\theta(1 + i)$ occurs with nonzero coefficient) has integer norm. So no new integers arise and $\mathfrak{O} = G'$.

Exercises

2.1 Which of the following complex numbers are algebraic? Which are algebraic *integers*?

(a) $355/113$

(b) $e^{2\pi i/23}$

(c) $e^{\pi i/23}$

(d) $\sqrt{17} + \sqrt{19}$

(e) $(1 + \sqrt{17})/(2\sqrt{-19})$

(f) $\sqrt{(1 + \sqrt{2})} + \sqrt{(1 - \sqrt{2})}$.

2.2 Express $\mathbf{Q}(\sqrt{3}, \sqrt[3]{5})$ in the form $\mathbf{Q}(\theta)$.

2.3 Find all monomorphisms $\mathbf{Q}(\sqrt[3]{7}) \to \mathbf{C}$.

2.4 Find the discriminant of $\mathbf{Q}(\sqrt{3}, \sqrt{5})$.

2.5 Let $K = \mathbf{Q}(\sqrt[4]{2})$. Find all monomorphisms $\sigma : K \to \mathbf{C}$ and the minimum polynomials (over $\mathbf{Q}$) and field polynomials (over K) of (i) $\sqrt[4]{2}$ (ii) $\sqrt{2}$ (iii) 2 (iv) $\sqrt{2}+1$. Compare with Theorem 3.5.

2.6 Complete Example 1 above by discussing the case $p = 3$.

2.7 Complete Example 2 above by discussing the case $p = 3$.

2.8 Compute integral bases and discriminants of
(a) $\mathbf{Q}(\sqrt{2}, \sqrt{3})$
(b) $\mathbf{Q}(\sqrt{2}, i)$
(c) $\mathbf{Q}(\sqrt[3]{2})$
(d) $\mathbf{Q}(\sqrt[4]{2})$.

2.9 Let $K = \mathbf{Q}(\theta)$ where $\theta \in \mathfrak{O}_K$. Among the elements

$$\frac{1}{d}(a_0 + \ldots + a_i\theta^i)$$

$(0 \neq a_i; a_0, \ldots, a_i \in \mathbf{Z})$, where d is the discriminant, pick one with minimal value of $|a_i|$ and call it x_i. Do this for $i = 1, \ldots, n = [K:\mathbf{Q}]$ show that $\{x_1, \ldots, x_n\}$ is an integral basis.

2.10 If $\alpha_1, \ldots, \alpha_n$ are $\mathbf{Q}$-linearly independent algebraic integers in $\mathbf{Q}(\theta)$, and if

$$\Delta[\alpha_1, \ldots, \alpha_n] = d$$

where d is the discriminant of $\mathbf{Q}(\theta)$, show that $\{\alpha_1, \ldots, \alpha_n\}$ is an integral basis for $\mathbf{Q}(\theta)$.

2.11 If $[K:\mathbf{Q}] = n$, $\alpha \in \mathbf{Q}$, show

$$N_K(\alpha) = \alpha^n,$$

$$T_K(\alpha) = n\alpha.$$

2.12 Give examples to show that for fixed α, $N_K(\alpha)$ and $T_K(\alpha)$ depend on K. (This is to emphasize that the norm and trace must always be defined in the context of a specific field K; there is no such thing as the norm or trace of α without a specified field.)

2.13 The norm and trace may be generalized by considering number fields $K \supseteq L$. Suppose $K = L(\theta)$ and $[K:L] = n$. Consider monomorphisms $\sigma : K \to \mathbf{C}$ such that $\sigma(x) = x$ for all $x \in L$. Show that there are precisely n such monomorphisms $\sigma_1, \ldots, \sigma_n$ and describe them. For $\alpha \in K$, define

$$N_{K/L}(\alpha) = \prod_{i=1}^{n} \sigma_i(\alpha),$$

$$T_{K/L}(\alpha) = \sum_{i=1}^{n} \sigma_i(\alpha).$$

(Compared with our earlier notation, we have $N_K = N_{K/\mathbf{Q}}$, $T_K = T_{K/\mathbf{Q}}$.) Prove that

$$N_{K/L}(\alpha_1\alpha_2) = N_{K/L}(\alpha_1)N_{K/L}(\alpha_2),$$

$$T_{K/L}(\alpha_1 + \alpha_2) = T_{K/L}(\alpha_1) + T_{K/L}(\alpha_2).$$

Let $K = \mathbf{Q}(\sqrt[4]{3})$, $L = \mathbf{Q}(\sqrt{3})$. Calculate $N_{K/L}(\alpha)$, $T_{K/L}(\alpha)$ for $\alpha = \sqrt[4]{3}$ and $\alpha = \sqrt[4]{3} + \sqrt{3}$.

2.14 For $K = \mathbf{Q}(\sqrt[4]{3})$, $L = \mathbf{Q}(\sqrt{3})$, calculate $N_{K/L}(\sqrt{3})$ and $N_{K/\mathbf{Q}}(\sqrt{3})$. Deduce that $N_{K/L}(\alpha)$ depends on K and L (provided that $\alpha \in K$). Do the same for $T_{K/L}$.

CHAPTER THREE

Quadratic and cyclotomic fields

In this chapter we investigate two special cases of number fields in the light of our previous work. The quadratic fields are those of degree 2, and are especially important in the study of quadratic forms. The cyclotomic fields are generated by pth roots of unity, and we consider only the case p prime; it is these which figure in Kummer's approach to Fermat's Last Theorem. We shall return to both types of field at later stages. For the moment we content ourselves with finding the rings of integers, integral bases, and discriminants.

3.1 Quadratic fields

A *quadratic field* is a number field K of degree 2 over $\mathbf{Q}$. Then $K = \mathbf{Q}(\theta)$ where θ is an algebraic integer, and θ is a zero of

$$t^2 - at + b \qquad (a, b \in \mathbf{Z}).$$

Thus

$$\theta = \frac{-a \pm \sqrt{(a^2 - 4b)}}{2}.$$

Let $a^2 - 4b = r^2 d$ where $r, d \in \mathbf{Z}$ and d is squarefree. (That this is always possible follows from prime factorization in $\mathbf{Z}$.) Then

$$\theta = \frac{-a \pm r\sqrt{d}}{2}$$

and so $\mathbf{Q}(\theta) = \mathbf{Q}(\sqrt{d})$. Hence we have proved:

Proposition 3.1. *The quadratic fields are precisely those of the form $\mathbf{Q}(\sqrt{d})$ for d a squarefree rational integer.* □

Next we determine the ring of integers of $\mathbf{Q}(\sqrt{d})$, for squarefree d. The answer, it turns out, depends on the arithmetic properties of d.

Theorem 3.2. *Let d be a squarefree rational integer. Then the integers of $\mathbf{Q}(\sqrt{d})$ are:*

(a) $\mathbf{Z}[\sqrt{d}]$ *if* $d \not\equiv 1 \pmod 4$,

(b) $\mathbf{Z}[\frac{1}{2} + \frac{1}{2}\sqrt{d}]$ *if* $d \equiv 1 \pmod 4$.

Proof. Every element $\alpha \in \mathbf{Q}(\sqrt{d})$ is of the form $\alpha = r + s\sqrt{d}$ for $r, s \in \mathbf{Q}$. Hence we may write

$$\alpha = \frac{a + b\sqrt{d}}{c}$$

where $a, b, c \in \mathbf{Z}$, $c > 0$, and no prime divides all of a, b, c. Now α is an integer if and only if the coefficients of the minimum polynomial

$$\left(t - \left(\frac{a + b\sqrt{d}}{c}\right)\right)\left(t - \left(\frac{a - b\sqrt{d}}{c}\right)\right)$$

are integers. Thus

$$\frac{a^2 - b^2 d}{c^2} \in \mathbf{Z}, \tag{1}$$

$$\frac{2a}{c} \in \mathbf{Z}. \tag{2}$$

If c and a have a common prime factor p then (1) implies that p divides b (since d is squarefree) which contradicts our previous assumption. Hence from (2) we have $c = 1$ or 2. If

$c = 1$ then α is an integer of K in any case, so we may concentrate on the case $c = 2$. Now a and b must both be odd, and $(a^2 - b^2d)/4 \in \mathbf{Z}$. Hence

$$a^2 - b^2d \equiv 0 \qquad (\text{mod } 4).$$

Now an odd number $2k + 1$ has square $4k^2 + 4k + 1 \equiv 1$ (mod 4), hence $a^2 \equiv 1 \equiv b^2$ (mod 4), and this implies $d \equiv 1$ (mod 4). Conversely, if $d \equiv 1$ (mod 4) then for odd a, b we have α an integer because (1) and (2) hold.

To sum up: if $d \not\equiv 1$ (mod 4) then $c = 1$ and so (a) holds; whereas if $d \equiv 1$ (mod 4) we can also have $c = 2$ and a, b odd, whence easily (b) holds. □

The monomorphisms $K \to \mathbf{C}$ are given by

$$\sigma_1(r + s\sqrt{d}) = r + s\sqrt{d},$$
$$\sigma_2(r + s\sqrt{d}) = r - s\sqrt{d}.$$

Hence we can compute discriminants:

Theorem 3.3. (a) *If* $d \not\equiv 1$ (mod 4) *then* $\mathbf{Q}(\sqrt{d})$ *has an integral basis of the form* $\{1, \sqrt{d}\}$ *and discriminant* $4d$. (b) *If* $d \equiv 1$ (mod 4) *then* $\mathbf{Q}(\sqrt{d})$ *has an integral basis of the form* $\{1, \frac{1}{2} + \frac{1}{2}\sqrt{d}\}$ *and discriminant* d.

Proof. The assertions regarding bases are clear from Theorem 3.2. Computing discriminants we work out:

$$\begin{vmatrix} 1 & \sqrt{d} \\ 1 & -\sqrt{d} \end{vmatrix}^2 = (-2\sqrt{d})^2 = 4d,$$

$$\begin{vmatrix} 1 & \frac{1}{2} + \frac{1}{2}\sqrt{d} \\ 1 & \frac{1}{2} - \frac{1}{2}\sqrt{d} \end{vmatrix}^2 = (-\sqrt{d})^2 = d. \qquad \square$$

Since the discriminants of isomorphic fields are equal, it follows that for distinct squarefree d the fields $\mathbf{Q}(\sqrt{d})$ are not isomorphic. This completes the classification of quadratic fields.

A special case, of historical interest as the first number

field to be studied as such, is the *Gaussian field* $\mathbf{Q}(\sqrt{-1})$. Since $-1 \not\equiv 1 \pmod 4$ the ring of integers is $\mathbf{Z}[\sqrt{-1}]$ (known as the ring of *Gaussian integers*) and the discriminant is -4.

Incidentally, these results show that Theorem 2.16 is not always applicable: an integral basis *can* have a discriminant which is not squarefree. For instance, the Gaussian integers themselves.

For future reference we note the norms and traces:

$$\begin{aligned} N(r + s\sqrt{d}) &= r^2 - ds^2, \\ T(r + s\sqrt{d}) &= 2r. \end{aligned}$$

We also note some useful terminology. A quadratic field $\mathbf{Q}(\sqrt{d})$ is said to be *real* if d is positive, *imaginary* if d is negative. (A real quadratic field contains only real numbers, an imaginary quadratic field contains proper complex numbers as well.)

3.2 Cyclotomic fields

A *cyclotomic field* is one of the form $\mathbf{Q}(\zeta)$ where $\zeta = e^{2\pi i/m}$ is a primitive complex mth root of unity. (The name means 'circle-cutting' and refers to the equal spacing of powers of ζ around the unit circle in the complex plane.) We shall consider only the case $m = p$, a prime number. Further, if $p = 2$ then $\zeta = -1$ so that $\mathbf{Q}(\zeta) = \mathbf{Q}$, hence we ignore this case and assume p odd.

Lemma 3.4. *The minimum polynomial of $\zeta = e^{2\pi i/p}$, p an odd prime, over $\mathbf{Q}$ is*

$$f(t) = t^{p-1} + t^{p-2} + \ldots + t + 1.$$

The degree of $\mathbf{Q}(\zeta)$ is $p - 1$.

Proof. We have

$$f(t) = \frac{t^p - 1}{t - 1}.$$

Since $\zeta - 1 \neq 0$ and $\zeta^p = 1$ it follows that $f(\zeta) = 0$, so all we need prove is that f is irreducible. This we do by a standard piece of trickery. We have

$$f(t+1) = \frac{(t+1)^p - 1}{t} = \sum_{r=1}^{p} \binom{p}{r} t^{r-1}.$$

Now the binomial coefficient $\binom{p}{r}$ is divisible by p if $1 \leqslant r \leqslant p-1$, and $\binom{p}{1} = p$ is not divisible by p^2. Hence by Eisenstein's criterion (Theorem 1.5) $f(t+1)$ is irreducible. Therefore $f(t)$ is irreducible, and is the minimum polynomial of ζ. Since $\partial f = p - 1$ we have $[\mathbf{Q}(\zeta):\mathbf{Q}] = p - 1$ by Theorem 1.8. □

The powers $\zeta, \zeta^2, \ldots, \zeta^{p-1}$ are also pth roots of unity, not equal to 1, and so by the same argument have $f(t)$ as minimum polynomial. Clearly

$$f(t) = (t-\zeta)(t-\zeta^2)\ldots(t-\zeta^{p-1}) \qquad (3)$$

and thus the conjugates of ζ are $\zeta, \zeta^2, \ldots, \zeta^{p-1}$. This means that the monomorphisms from $\mathbf{Q}(\zeta)$ to $\mathbf{C}$ are given by

$$\sigma_i(\zeta) = \zeta^i \qquad (1 \leqslant i \leqslant p-1).$$

Because the minimum polynomial $f(t)$ has degree $p - 1$, a basis for $\mathbf{Q}(\zeta)$ over $\mathbf{Q}$ is $1, \zeta, \ldots, \zeta^{p-2}$, so for a general element

$$\alpha = a_0 + a_1\zeta + \ldots + a_{p-2}\zeta^{p-2} \qquad (a_i \in \mathbf{Q})$$

we have

$$\sigma_i(a_0 + \zeta + \ldots + a_{p-2}\zeta^{p-2}) = a_0 + \zeta^i + \ldots + a_{p-2}\zeta^{i(p-2)}.$$

From this formula the norm and trace may be calculated using the basic definitions

$$\mathrm{N}(\alpha) = \prod_{i=1}^{p-1} \sigma_i(\alpha),$$

$$T(\alpha) = \sum_{i=1}^{p-1} \sigma_i(\alpha).$$

In particular

$$N(\zeta) = \zeta.\zeta^2 \ldots \zeta^{p-1}.$$

Now ζ and ζ^i $(1 \leqslant i \leqslant p-1)$ are conjugates, so have the same norm, which can be calculated by putting $t = 0$ in (3) to give

$$N(\zeta) = N(\zeta^i) = (-1)^{p-1}$$

and since p is odd,

$$N(\zeta^i) = 1 \qquad (1 \leqslant i \leqslant p-1). \tag{4}$$

The trace of ζ^i can be found by a similar argument. We have

$$T(\zeta^i) = T(\zeta) = \zeta + \zeta^2 + \ldots + \zeta^{p-1},$$

and using the fact that

$$f(\zeta) = 1 + \zeta + \ldots + \zeta^{p-1} = 0$$

we find

$$T(\zeta^i) = -1 \qquad (1 \leqslant i \leqslant p-1). \tag{5}$$

For $a \in \mathbf{Q}$ we trivially have

$$N(a) = a^{p-1}$$
$$T(a) = (p-1)a.$$

Since $\zeta^p = 1$, we can use these formulae to extend (4) and (5) to

$$N(\zeta^s) = 1 \qquad \text{for all } s \in \mathbf{Z} \tag{6}$$

and

$$T(\zeta^s) = \begin{cases} -1 & \text{if } s \not\equiv 0 \pmod{p} \\ p-1 & \text{if } s \equiv 0 \pmod{p}. \end{cases} \tag{7}$$

For a general element of $\mathbf{Q}(\zeta)$, the trace is easily calculated:

$$T\left(\sum_{i=0}^{p-2} a_i\zeta^i\right) = \sum_{i=0}^{p-2} T(a_i\zeta^i)$$

$$= \mathrm{T}(a_0) + \sum_{i=1}^{p-2} \mathrm{T}(a_i\zeta^i)$$

$$= (p-1)a_0 - \sum_{i=0}^{p-2} a_i$$

and so

$$\mathrm{T}\left(\sum_{i=0}^{p-2} a_i\zeta^i\right) = pa_0 - \sum_{i=0}^{p-2} a_i. \tag{8}$$

The norm is more complicated in general, but a useful special case is

$$\mathrm{N}(1-\zeta) = \prod_{i=1}^{p-1} (1-\zeta^i)$$

which can be calculated by putting $t = 1$ in (3) to obtain

$$\prod_{i=1}^{p-1} (1-\zeta^i) = p, \tag{9}$$

so

$$\mathrm{N}(1-\zeta) = p. \tag{10}$$

We can put these computations to good use, first by showing that the integers of $\mathbf{Q}(\zeta)$ are what one naively might expect:

Theorem 3.5. *The ring* $\mathfrak{O}$ *of integers of* $\mathbf{Q}(\zeta)$ *is* $\mathbf{Z}[\zeta]$.

Proof. Suppose $\alpha = a_0 + a_1\zeta + \ldots + a_{p-2}\zeta^{p-2}$ is an integer in $\mathbf{Q}(\zeta)$. We must demonstrate that the rational numbers a_i are actually rational integers.

For $0 \leqslant k \leqslant p-2$ the element

$$\alpha\zeta^{-k} - \alpha\zeta$$

is an integer, so its trace is a rational integer. But

$$\mathrm{T}(\alpha\zeta^{-k} - \alpha\zeta)$$

$$= \mathrm{T}(a_0\zeta^{-k} + \ldots + a_k + \ldots + a_{p-2}\zeta^{p-k-2} - a_0\zeta - \ldots - a_{p-2}\zeta^{p-1})$$

$$= pa_k - (a_0 + \ldots + a_{p-2}) - (-a_0 - \ldots - a_{p-2})$$

$$= pa_k.$$

Hence $b_k = pa_k$ is a rational integer.

Put $\lambda = 1 - \zeta$. Then

$$\begin{aligned} p\alpha &= b_0 + b_1\zeta + \ldots + b_{p-2}\zeta^{p-2} \\ &= c_0 + c_1\lambda + \ldots + c_{p-2}\lambda^{p-2} \end{aligned} \tag{11}$$

where (substituting $\zeta = 1 - \lambda$ and expanding)

$$c_i = \sum_{j=i}^{p-2} (-1)^i \binom{j}{i} b_j \in \mathbf{Z}.$$

Since $\lambda = 1 - \zeta$ we also have, symmetrically,

$$b_i = \sum_{j=i}^{p-2} (-1)^i \binom{j}{i} c_j. \tag{12}$$

We claim that all c_i are divisible by p. Proceeding by induction, we may assume this for all c_i with $i \leqslant k - 1$, where $0 \leqslant k \leqslant p - 2$. Since $c_0 = b_0 + \ldots + b_{p-2} = p(-T(\alpha) + b_0)$, we have $p|c_0$, so it is true for $k = 0$. Now by (9)

$$p = \prod_{i=1}^{p-1} (1 - \zeta^i)$$

$$= (1 - \zeta)^{p-1} \prod_{i=1}^{p-1} (1 + \zeta + \ldots + \zeta^{i-1})$$

$$= \lambda^{p-1}\kappa \tag{13}$$

where $\kappa \in \mathbf{Z}[\zeta] \subseteq \mathfrak{O}$. Consider (11) as a congruence modulo the ideal $\langle\lambda^{k+1}\rangle$ of $\mathfrak{O}$. By (13) we have

$$p \equiv 0 \qquad (\text{mod } \langle\lambda^{k+1}\rangle)$$

and so the left-hand side of (11), and the terms up to $c_{k-1}\lambda^{k-1}$, vanish; further the terms from $c_{k+1}\lambda^{k+1}$ onwards are multiples of λ^{k+1} so also vanish. There remains:

$$c_k\lambda^k \equiv 0 \qquad (\text{mod } \langle\lambda^{k+1}\rangle).$$

This is equivalent to

$$c_k\lambda^k = \mu\lambda^{k+1}$$

for some $\mu \in \mathfrak{O}$, from which we obtain

$$c_k = \mu\lambda.$$

Taking norms we get

$$c_k^{p-1} = \mathrm{N}(c_k) = \mathrm{N}(\mu)\mathrm{N}(\lambda) = p\mathrm{N}(\mu),$$

since $\mathrm{N}(\lambda) = p$ by (10). Hence $p \mid c_k^{p-1}$, so $p \mid c_k$. Hence by induction $p \mid c_k$ for all k, and then (11) shows that $p \mid b_k$ for all k. Therefore $a_k \in \mathbf{Z}$ for all k and the theorem is proved. □

Now we can compute the discriminant.

Theorem 3.6. *The discriminant of* $\mathbf{Q}(\zeta)$, *where* $\zeta = e^{2\pi i/p}$ *and* p *is an odd prime, is*

$$(-1)^{(p-1)/2}.p^{p-2}.$$

Proof. By Theorem 3.5 an integral basis is $\{1, \zeta, \ldots, \zeta^{p-2}\}$. Hence by Proposition 2.17 the discriminant is equal to

$$(-1)^{(p-1)(p-2)/2}.\mathrm{N}(\mathrm{D}f(\zeta))$$

with $f(t)$ as above. Since p is odd the first factor reduces to $(-1)^{(p-1)/2}$. To evaluate the second, we have

$$f(t) = \frac{t^p - 1}{t - 1},$$

so that

$$\mathrm{D}f(t) = \frac{(t-1)pt^{p-1} - (t^p - 1)}{(t-1)^2}$$

whence

$$\mathrm{D}f(\zeta) = \frac{-p\zeta^{p-1}}{\lambda}$$

where $\lambda = 1 - \zeta$ as before. Hence

$$\mathrm{N}(\mathrm{D}f(\zeta)) = \frac{\mathrm{N}(p)\mathrm{N}(\zeta)^{p-1}}{\mathrm{N}(\lambda)}$$

$$= \frac{(-p)^{p-1}1^{p-1}}{p}$$

$$= p^{p-2}. \qquad \square$$

The case $p = 3$ deserves special mention, for $\mathbf{Q}(\zeta)$ has degree $p - 1 = 2$, so it is a quadratic field. Since

$$e^{2\pi i/3} = \frac{-1+\sqrt{-3}}{2}$$

it is equal to $\mathbf{Q}(\sqrt{-3})$. As a check on our discriminant calculations: Theorem 3.3 gives -3 (since $-3 \equiv 1 \pmod 4$), and Theorem 3.6 gives $(-1)^{2/2}3^1 = -3$ as well.

Exercises

3.1 Find integral bases and discriminants for:
(a) $\mathbf{Q}(\sqrt{3})$
(b) $\mathbf{Q}(\sqrt{-7})$
(c) $\mathbf{Q}(\sqrt{11})$
(d) $\mathbf{Q}(\sqrt{-11})$
(e) $\mathbf{Q}(\sqrt{6})$
(f) $\mathbf{Q}(\sqrt{-6})$

3.2 Let $K = \mathbf{Q}(\zeta)$ where $\zeta = e^{2\pi i/5}$. Calculate $N_K(\alpha)$ and $T_K(\alpha)$ for the following values of α:
(i) ζ^2 (ii) $\zeta + \zeta^2$ (iii) $1 + \zeta + \zeta^2 + \zeta^3 + \zeta^4$.

3.3 Let $K = \mathbf{Q}(\zeta)$ where $\zeta = e^{2\pi i/p}$ for a rational prime p. In the ring of integers $\mathbf{Z}[\zeta]$, show that $\alpha \in \mathbf{Z}[\zeta]$ is a unit if and only if $N_K(\alpha) = \pm 1$.

3.4 If $\zeta = e^{2\pi i/3}$, $K = \mathbf{Q}(\zeta)$, prove that the norm of $\alpha \in \mathbf{Z}[\zeta]$ is of the form $\frac{1}{4}(a^2 + 3b^2)$ where a, b are rational integers which are either both even or both odd. Using the result of question 3.3, deduce that there are precisely six units in $\mathbf{Z}[\zeta]$ and find them all.

3.5 If $\zeta = e^{2\pi i/5}$, $K = \mathbf{Q}(\zeta)$, prove that the norm of

$\alpha \in \mathbf{Z}[\zeta]$ is of the form $\frac{1}{4}(a^2 - 5b^2)$ where a, b are rational integers. (Hint: in calculating $\mathrm{N}(\alpha)$, first calculate $\sigma_1(\alpha)\sigma_4(\alpha)$ where $\sigma_i(\zeta) = \zeta^i$. Show that this is of the form $q + r\theta + s\phi$ where q, r, s are rational integers, $\theta = \zeta + \zeta^4$, $\phi = \zeta^2 + \zeta^3$. In the same way, establish $\sigma_2(\alpha)\sigma_3(\alpha) = q + s\theta + r\phi$.) Using Question 3.3, prove that $\mathbf{Z}[\zeta]$ has an infinite number of units.

3.6 Let $\zeta = e^{2\pi i/5}$. For $K = \mathbf{Q}(\zeta)$, use the formula $\mathrm{N}_K(a + b\zeta) = (a^5 + b^5)/(a + b)$ to calculate the following norms:
(i) $\mathrm{N}_K(\zeta + 2)$ (ii) $\mathrm{N}_K(\zeta - 2)$ (iii) $\mathrm{N}_K(\zeta + 3)$.
Using the fact that if $\alpha\beta = \gamma$, then $\mathrm{N}_K(\alpha)\mathrm{N}_K(\beta) = \mathrm{N}_K(\gamma)$, deduce that $\zeta + 2$, $\zeta - 2$, $\zeta + 3$ have no proper factors (i.e. factors which are not units) in $\mathbf{Z}[\zeta]$.
Factorize 11, 31, 61 in $\mathbf{Z}[\zeta]$.

3.7 If $\zeta = e^{2\pi i/5}$, as in Question 3.6, calculate
(i) $\mathrm{N}_K(\zeta + 4)$ (ii) $\mathrm{N}_K(\zeta - 3)$.
Deduce that any proper factors of $\zeta + 4$ in $\mathbf{Z}[\zeta]$ have norm 5 or 41. Given $\zeta - 1$ is a factor of $\zeta + 4$, find another factor. Verify $\zeta - 3$ is a unit times $(\zeta^2 + 2)^2$ in $\mathbf{Z}[\zeta]$.

3.8 Show that the multiplicative group of non-zero elements of $\mathbf{Z}_7$ is cyclic with generator the residue class of 3. If $\zeta = e^{2\pi i/7}$, define the monomorphism $\sigma : \mathbf{Q}(\zeta) \to \mathbf{C}$ by $\sigma(\zeta) = \zeta^3$. Show that all other monomorphisms from $\mathbf{Q}(\zeta)$ to $\mathbf{C}$ are of the form σ^i ($1 \leqslant i \leqslant 6$) where $\sigma^6 = 1$. For any $\alpha \in \mathbf{Q}(\zeta)$, define $c(\alpha) = \alpha\sigma^2(\alpha)\sigma^4(\alpha)$, and show $\mathrm{N}(\alpha) = c(\alpha) . \sigma c(\alpha)$. Demonstrate that $c(\alpha) = \sigma^2 c(\alpha) = \sigma^4 c(\alpha)$. Using the relation $1 + \zeta + \ldots + \zeta^6 = 0$, show that every element $\alpha \in \mathbf{Q}(\zeta)$ can be written uniquely as $\Sigma_{i=1}^{6} a_i\zeta^i$ ($a_i \in \mathbf{Q}$). Deduce that $c(\alpha) = a_1\theta_1 + a_3\theta_2$ where $\theta_1 = \zeta + \zeta^2 + \zeta^4$, $\theta_2 = \zeta^3 + \zeta^5 + \zeta^6$. Show $\theta_1 + \theta_2 = -1$ and calculate $\theta_1\theta_2$. Verify that $c(\alpha)$ may be written in the form $b_0 + b_1\theta_1$ where $b_0, b_1 \in \mathbf{Q}$, and show

$\sigma c(\alpha) = b_0 + b_1\theta_2$. Deduce

$$N(\alpha) = b_0^2 - b_0 b_1 + 2b_1^2.$$

Now calculate $N(\zeta + 5\zeta^6)$.

3.9 Suppose p is a rational prime and $\zeta = e^{2\pi i/p}$. Given that the group of non-zero elements of $\mathbf{Z}_p$ is cyclic (see Appendix 1, Proposition 6 for a proof) show that there exists a monomorphism $\sigma : \mathbf{Q}(\zeta) \to \mathbf{C}$ such that σ^{p-1} is the identity and all monomorphisms from $\mathbf{Q}(\zeta)$ to $\mathbf{C}$ are of the form σ^i $(1 \leqslant i \leqslant p-1)$. If $p - 1 = kr$, define $c_k(\alpha) = \alpha\sigma^r(\alpha)\sigma^{2r}(\alpha) \ldots \sigma^{(k-1)r}(\alpha)$. Show $N(\alpha) = c_k(\alpha) . \sigma c_k(\alpha) \ldots \sigma^{r-1} c_k(\alpha)$. Prove every element of $\mathbf{Q}(\zeta)$ is uniquely of the form $\Sigma_{i=1}^{p-1} a_i\zeta^i$, and by demonstrating that $\sigma^r(c_k(\alpha)) = c_k(\alpha)$, deduce that $c_k(\alpha) = b_1\eta_1 + \ldots + b_k\eta_k$, where

$$\eta_1 = \zeta + \sigma^r(\zeta) + \sigma^{2r}(\zeta) + \ldots + \sigma^{(k-1)r}(\zeta)$$

and $\eta_{i+1} = \sigma^i(\eta_1)$.
Interpret these results in the case $p = 5$, $k = r = 2$, by showing that the residue class of 2 is a generator of the multiplicative group of non-zero elements of $\mathbf{Z}_5$. Demonstrate that $c_2(\alpha)$ is of the form $b_1\eta_1 + b_2\eta_2$ where $\eta_1 = \zeta + \zeta^4$, $\eta_2 = \zeta^2 + \zeta^3$.
Calculate the norms of the following elements in $\mathbf{Q}(\zeta)$:
(i) $\zeta + 2\zeta^2$ (ii) $\zeta + \zeta^4$ (iii) $15\zeta + 15\zeta^4$
(iv) $\zeta + \zeta^2 + \zeta^3 + \zeta^4$.

3.10 In $\mathbf{Z}[\sqrt{-5}]$, prove 6 factorizes in two ways as

$$6 = 2.3 = (1 + \sqrt{-5})(1 - \sqrt{-5})$$

Verify that 2, 3, $1 + \sqrt{-5}$, $1 - \sqrt{-5}$ have no proper factors in $\mathbf{Z}[\sqrt{-5}]$. (Hint: Take norms and note that if γ factorizes as $\gamma = \alpha\beta$, then $N(\gamma) = N(\alpha)N(\beta)$ is a factorization of rational integers.) Deduce that it is possible in $\mathbf{Z}[\sqrt{-5}]$ for 2 to have no proper factors, yet 2 divides a product $\alpha\beta$ without dividing either α or β.

CHAPTER FOUR

Factorization into irreducibles

We now generalize the theory of prime factorization in **Z** to an arbitrary integral domain D, and in particular to the ring of integers $\mathfrak{O}$ of a number field. In **Z** we can factorize into prime numbers and obtain a factorization which is unique except for the order of factors and the presence of units ± 1. Such a notion of unique factorization does not carry over to all rings of integers, but it does hold in some cases. As we shall see this caused a great deal of confusion in the history of the subject. The nub of the problem turned out to be the definition of a prime. In **Z** a prime number has two basic properties:

(1) $m \mid p$ implies $m = \pm p$ or ± 1,

(2) $p \mid mn$ implies $p \mid m$ or $p \mid n$.

Either of these will do as the definition of a prime number in **Z**, and we usually take the former. In an arbitrary domain, it turns out that property (2) is what is required for uniqueness of factorization and in general (2) does not follow from (1). Property (1) is simply the definition of an *irreducible* element in **Z**. We will reserve the term *prime* for an element which satisfies (2) and is neither zero nor a unit. A prime is always irreducible, but not vice versa.

When we factorize an element x in a domain D, it is natural to seek proper factors $x = ab$; then if these are further reducible, to factorize them and so on, seeking a factorization

$$x = p_1 \dots p_m$$

where all the factors are irreducible. In general this may not be possible, but it is so in a ring $\mathfrak{O}$ of integers in any number field. (To prove this we introduce the notion of a noetherian ring, and show that factorization is always possible if the domain D is noetherian; we then demonstrate that $\mathfrak{O}$ is noetherian.) Even though factorization into irreducibles is always possible in $\mathfrak{O}$, we give an extensive list of examples where such a factorization is not unique. In other cases, however, the existence of a generalized version of the division algorithm (which we term a 'Euclidean function') implies that every irreducible is prime. We see that factorization into primes is unique, so some rings $\mathfrak{O}$ possess unique factorization. In particular we characterize such $\mathfrak{O}$ for the fields $\mathbf{Q}(\sqrt{d})$ with d a negative rational integer: there are exactly five of them, corresponding to $d = -1, -2, -3, -7, -11$. We also prove the existence of a Euclidean function for some fields $\mathbf{Q}(\sqrt{d})$ with d positive. In later chapters we shall see that $\mathfrak{O}$ may have unique factorization without possessing a Euclidean function.

To begin the chapter we consider a little history, and look at an example of the intuitive use of unique factorization, to motivate the ideas.

4.1 Historical background

In the 18th century and the first part of the 19th there were varying standards of rigour in number theory. For example Euler, the prolific mathematician of the 18th century, was concerned with obtaining results and would, on occasion, use intuitive methods of proof which the hindsight of history has shown to be incorrect. For instance, in his famous textbook on algebra, he made several elegant applications of unique factorization to 'prove' number theoretic results in cases where unique factorization was false. Gauss, on the other hand, found it necessary to demonstrate rigorously that the so-called 'Gaussian integers' $\mathbf{Z}[i]$ did factorize uniquely. In 1847, Lamé announced to a meeting of the Paris Academy that he had solved Fermat's Last Theorem, but his proof was seen to depend on uniqueness of factorization and was shown to be inadequate. Kummer had, in fact, published a paper three years earlier that demonstrated the failure of unique factorization for cyclotomic integers, thus destroying Lamé's proof, but his publication was in an obscure journal and went unnoticed at the time.

Eisenstein put his finger on the property that characterizes unique factorization in a letter of 1844 which translates

'. . . if one had the theorem which states that the product of two complex numbers can be divisible by a prime number only when one of the factors is – which seems completely obvious – then one would have the whole theory at a single blow; but this theorem is totally false . . . '.

By 'the whole theory' he was referring to consequences of unique factorization (in particular Fermat's Last Theorem).

In Eisenstein's letter, 'prime' meant definition (1) of this chapter, and his comment translated into the terminology of this book is 'if every irreducible is prime, then unique factorization holds'. It is also clear from his comment that he knew of instances of irreducibles which were not prime, which gave rise to non-uniqueness of factorization. All this must have seemed very confusing to average 19th century mathematicians who were used to using intuitive ideas about factorization to demonstrate results. To give the reader an idea of what it was like, before we explain the theory of unique factorization, we give a concocted proof of a statement of Fermat which uses this intuitive language. Fermat's proof has not survived, and we are not suggesting that it resembled our faulty but instructive attempt below. Indeed it is not hard to reconstruct a rigorous proof using ideas that were known to Fermat.

A statement of Fermat. *The equation* $y^2 + 2 = x^3$ *has only the integer solutions* $y = \pm 5, x = 3$.

Intuitive 'proof'. Clearly y cannot be even, for then the right-hand side would be divisible by 8, but the left-hand side only by 2. We now factorize in the ring $\mathbf{Z}[\sqrt{-2}]$, consisting of all $a + b\sqrt{-2}$ for $a, b \in \mathbf{Z}$, to give

$$(y + \sqrt{-2})(y - \sqrt{-2}) = x^3.$$

A common factor $c + d\sqrt{-2}$ of $y + \sqrt{-2}$ and $y - \sqrt{-2}$ would also divide their sum $2y$ and their difference $2\sqrt{-2}$. Taking norms,

$$c^2 + 2d^2 \mid 4y^2, \qquad c^2 + 2d^2 \mid 8,$$

hence $c^2 + 2d^2 \mid 4$. The only solutions of this relation are $c = \pm 1, d = 0$, or $c = 0, d = \pm 1$, or $c = \pm 2, d = 0$. None of these give proper factors of $y + \sqrt{-2}$, so $y + \sqrt{-2}$ and $y - \sqrt{-2}$ are coprime. Now the product of two coprime numbers is a cube only when each is a cube, so

$$y + \sqrt{-2} = (a + b\sqrt{-2})^3,$$

and comparing coefficients of $\sqrt{-2}$,

$$1 = b(3a^2 - 2b^2)$$

for which the only solutions are $b = 1, a = \pm 1$. Then $x = 3$, $y = \pm 5$. □

The flaw in this intuitive 'proof', as it stands, is that we are carrying over the language of factorization of integers to factorization in $\mathbf{Z}[\sqrt{-2}]$ without checking that the usual properties actually hold in $\mathbf{Z}[\sqrt{-2}]$. In this chapter we develop the appropriate theory and investigate when it generalizes to a ring of integers in a number field.

4.2 Trivial factorizations

If u is a unit in a ring R, then any element $x \in R$ can be trivially factorized as

$$x = uy$$

where $y = u^{-1}x$. An element y is called an *associate* of x if

$x = uy$ for a unit u. Recall that a factorization of $x \in R$, $x = yz$ is said to be proper if neither of y or z are units. If a factorization is not proper (in which case we shall call it *trivial*) then it is clear that one of the factors is a unit and the other is an associate of x. Before going on to proper factorizations we therefore look at elementary properties of units and associates. We denote the set of units in a ring R by $U(R)$.

Proposition 4.1. *The units $U(R)$ of a ring R form a group under multiplication.* □

Examples.
1. $R = \mathbf{Q}$. The units are $U(\mathbf{Q}) = \mathbf{Q} \setminus \{0\}$, which is an infinite group.
2. $R = \mathbf{Z}$. The units are ± 1, so $U(\mathbf{Z})$ is cyclic of order 2.
3. $R = \mathbf{Z}[i]$, the Gaussian integers, $a + ib$ $(a, b \in \mathbf{Z})$. The element $a + ib$ is a unit if and only if there exists $c + id$ $(c, d \in \mathbf{Z})$ such that

$$(a + ib)(c + id) = 1.$$

This implies $ac - bd = 1, ad + bc = 0$, whence $c = a/(a^2 + b^2), d = -b/(a^2 + b^2)$. These have integer solutions only when $a^2 + b^2 = 1$, so $a = \pm 1, b = 0$, or $a = 0, b = \pm 1$. Hence the units are $\{1, -1, i, -i\}$ and $U(R)$ is cyclic of order 4.

By using norms, we can extend the results of Example 3 to the more general case of the units in the ring of integers of $\mathbf{Q}(\sqrt{d})$ for d negative and squarefree:

Proposition 4.2. *The group of units U of the integers in $\mathbf{Q}(\sqrt{d})$ where d is negative and squarefree is as follows:*
(a) For $d = -1$, $U = \{\pm 1, \pm i\}$.
(b) For $d = -3$, $U = \{\pm 1, \pm \omega, \pm \omega^2\}$ where $\omega = e^{2\pi i/3}$.
(c) For all other $d < 0$, $U = \{\pm 1\}$.

Proof. Suppose α is a unit in the ring of integers of $\mathbf{Q}(\sqrt{d})$ with inverse β; then $\alpha\beta = 1$, so taking norms

$$N(\alpha)N(\beta) = 1.$$

But $N(\alpha)$, $N(\beta)$ are rational integers, so $N(\alpha) = \pm 1$. Writing $\alpha = a + b\sqrt{d}$ ($a, b \in \mathbf{Q}$), then we see that $N(\alpha) = a^2 - db^2$ is positive (for negative d), so $N(\alpha) = +1$. Hence we are reduced to solving the equation

$$a^2 - db^2 = 1.$$

If $a, b \in \mathbf{Z}$, then for $d = -1$ this reduces to

$$a^2 + b^2 = 1$$

which has the solutions $a = \pm 1, b = 0$, or $a = 0, b = \pm 1$, already found in Example 3. This gives (a). For $d < -3$ we immediately conclude that $b = 0$ (otherwise $a^2 - db^2$ would exceed 1), so the only rational integer solutions are $a = \pm 1$, $b = 0$. If $d \not\equiv 1 \pmod 4$, then $a, b \in \mathbf{Z}$, so the only solutions are those discovered. For $d \equiv 1 \pmod 4$, however, we must also consider the additional possibility $a = A/2$, $b = B/2$ where both A and B are odd rational integers. In this case

$$A^2 - dB^2 = 4.$$

For $d < -3$, we deduce $B = 0$ and there are no additional solutions. This completes (c). For $d = -3$, we find additional solutions $A = \pm 1, B = \pm 1$. The case $A = 1, B = 1$ gives

$$\alpha = \tfrac{1}{2}(-1 + \sqrt{-3}) = e^{2\pi i/3}$$

which we have denoted by ω. The other three cases give $-\omega$, $\omega^2, -\omega^2$. These allied with the solutions already found give (b). □

The general case of units in a ring of integers in a number field will be postponed until Chapter 11. We now return to simple properties of units and associates.

It follows from Proposition 4.1 that 'being associates' is an equivalence relation on R. The only associate of 0 is 0 itself. Recall that a non-unit $x \in R$ is called an irreducible if it has no proper factors. The zero element $0 = 0.0$ has factors, neither of which is a unit, so in particular an irreducible is non-zero. We now list a few elementary properties of units, associates and irreducibles. To prove some of these we shall require the cancellation law, so we must take the ring to be an integral domain.

Proposition 4.3. *For a domain D,*

(a) *x is a unit if and only if $x \mid 1$,*

(b) *Any two units are associates and any associate of a unit is a unit,*

(c) *x, y are associates if and only if $x \mid y$ and $y \mid x$,*

(d) *x is irreducible if and only if every divisor of x is an associate of x or a unit*

(e) *an associate of an irreducible is irreducible.*

Proof. Most of these follow straight from the definitions. We prove (c) which requires the cancellation law. Suppose $x \mid y$ and $y \mid x$; then there exist $a, b \in D$ such that $y = ax$, $x = by$. Substituting, we find

$$x = bax.$$

Now either $x = 0$, in which case $y = 0$ also and they are associates, or $x \neq 0$ and we cancel x to find

$$1 = ba,$$

so a and b are units. Hence x, y are associates. The converse is trivial. □

Some of these ideas may be usefully expressed in terms of ideals:

Proposition 4.4. *If D is a domain and x, y are non-zero elements of D then*

(a) *$x \mid y$ if and only if $\langle x \rangle \supseteq \langle y \rangle$,*

(b) *x and y are associates if and only if $\langle x \rangle = \langle y \rangle$,*

(c) *x is a unit if and only if $\langle x \rangle = D$,*

(d) *x is irreducible if and only if $\langle x \rangle$ is maximal among the proper principal ideals of D.*

Proof. (a) If $x \mid y$ then $y = zx \in \langle x \rangle$ for some $z \in D$, hence $\langle y \rangle \subseteq \langle x \rangle$. Conversely, if $\langle y \rangle \subseteq \langle x \rangle$ then $y \in \langle x \rangle$, so $y = zx$ for some $z \in D$.

(b) is immediate from (a).

(c) If x is a unit then $xv = 1$ for some $v \in D$, hence for any $y \in D$ we have $y = xvy \in \langle x \rangle$ and $D = \langle x \rangle$. If $D = \langle x \rangle$ then since $1 \in D$, $1 = zx$ for some $z \in D$ and x is a unit.

(d) Suppose x is irreducible, with $\langle x \rangle \subsetneqq \langle y \rangle \subsetneqq D$. Then $y \mid x$ but is neither a unit, nor an associate of x, contradicting 4.3 (d). Conversely, if no such y exists, then every divisor of x is either a unit or an associate, so x is irreducible. □

4.3 Factorization into irreducibles

In a domain D, if a non-unit is reducible, we can write it as

$$x = ab.$$

If either of a or b is reducible, we can express it as a product of proper factors; then carry on the process, seeking to write

$$x = p_1 p_2 \dots p_m$$

where each p_i is irreducible. We say that *factorization into irreducibles* is *possible* in D if every $x \in D$, not a unit nor zero, is a product of a finite number of irreducibles. In general such a factorization may not be possible; and an example is ready to hand, namely the ring $\mathbf{B}$ of all algebraic integers. For if α is not zero or a unit, neither is $\sqrt{\alpha}$. Since $\alpha = \sqrt{\alpha}.\sqrt{\alpha}$ and $\sqrt{\alpha}$ is an integer, it follows that α is not irreducible. Thus $\mathbf{B}$ has no irreducibles at all, but it does have non-zero non-units, so factorization into irreducibles is not possible.

This trouble does not arise in the ring $\mathfrak{O}$ of integers of a number field (which is another reason why we concentrate on such rings instead of the whole of $\mathbf{B}$). We will prove the possibility of factorization in $\mathfrak{O}$ by introducing a more general notion which makes the arguments involved more transparent. We define a domain D to be *noetherian* if every ideal in D is finitely generated. The adjective commemorates Emmy Noether (1882–1935) who introduced the concept.

Having demonstrated the possibility of factorization in any noetherian ring, we will show that $\mathfrak{O}$ is noetherian, so factorization is possible here also.

Two useful properties, which we shall see are each equivalent to the noetherian condition are:

The ascending chain condition. Given an ascending chain of ideals:

$$I_0 \subseteq I_1 \subseteq \ldots \subseteq I_n \subseteq \ldots \tag{1}$$

then there exists some N for which $I_n = I_N$ for all $n \geqslant N$. That is, every ascending chain *stops*.

The maximal condition. Every non-empty set of ideals has a maximal element, that is an element which is not properly contained in every other element.

We remark that this maximal element need not contain *all* the other ideals in the given set: we require only that there is no other element in the set that contains *it*.

Proposition 4.5. *The following conditions are equivalent for an integral domain D:*

(a) *D is noetherian,*
(b) *D satisfies the ascending chain condition,*
(c) *D satisfies the maximal condition.*

Proof. Assume (a). Consider an ascending chain as in (1). Let $I = \cup_{n=1}^{\infty} I_n$. Then I is an ideal, so finitely generated: say $I = \langle x_1, \ldots, x_m \rangle$. Each x_i belongs to some $I_{n(i)}$. If we let $N = \max_i n(i)$, then we have $I = I_N$ and it follows that $I_n = I_N$ for all $n \geqslant N$, proving (b).

Now suppose (b) and consider a non-empty set S of ideals. Suppose for a contradiction that S does not have a maximal element. Pick $I_0 \in S$. Since I_0 is not maximal we can pick $I_1 \in S$ with $I_0 \subsetneqq I_1$. Inductively, having found I_n, since this is not maximal, we can pick $I_{n+1} \in S$ with $I_n \subsetneqq I_{n+1}$. But now we have an ascending chain which does not stop, which is a contradiction. So (b) implies (c). (The reader who wishes to may ponder the use of the axiom of choice in this proof.)

Finally, suppose (c). Let I be any ideal, and let S be the set of all finitely generated ideals contained in I. Then $\{0\} \in S$, so S is non-empty and thus has a maximal element J. If $J \neq I$, pick $x \in I \backslash J$. Then $\langle J, x \rangle$ is finitely generated and strictly larger than J, a contradiction. Hence $J = I$ and I is finitely generated. □

Theorem 4.6. *If a domain D is noetherian, then factorization into irreducibles is possible in D.*

Proof. Suppose that D is noetherian, but there exists a non-unit $x \neq 0$ in D which cannot be expressed as a product of a finite number of irreducibles. Choose x so that $\langle x \rangle$ is maximal subject to these conditions on x, which is possible by the maximal condition. By its definition, this x cannot be irreducible, so $x = yz$ where y and z are not units. Then $\langle y \rangle \supseteq \langle x \rangle$ by 4.4(a). If $\langle y \rangle = \langle x \rangle$ then x and y are associates by 4.4(b) and this is not the case because it implies that z is a unit. So $\langle y \rangle \supsetneq \langle x \rangle$, and similarly $\langle z \rangle \supsetneq \langle x \rangle$. By maximality of $\langle x \rangle$, we must have

$$y = p_1 \dots p_r,$$

$$z = q_1 \dots q_s,$$

where each p_i and q_j is irreducible. Multiplying these together expresses x as a product of irreducibles, a contradiction. Hence the assumption that there existed a non-unit $\neq 0$ which is not a finite product of irreducibles is false, and factorization into irreducibles is always possible. □

We are now in business because:

Theorem 4.7. *The ring of integers $\mathfrak{O}$ in a number field K is noetherian.*

Proof. We prove that every ideal I of $\mathfrak{O}$ is finitely generated. Now $(\mathfrak{O}, +)$ is free abelian of rank n equal to the degree of

K by Theorem 2.15. Hence $(I, +)$ is free abelian of rank $s \leqslant n$ by Theorem 1.12. If $\{x_1, \ldots, x_s\}$ is a $\mathbf{Z}$-basis for $(I, +)$, then clearly $\langle x_1, \ldots, x_s \rangle = I$, so I is finitely generated and $\mathfrak{O}$ is noetherian. □

Corollary 4.8. *Factorization into irreducibles is possible in* $\mathfrak{O}$. □

To get very far in the theory, we need some easy ways of detecting units and irreducibles in $\mathfrak{O}$. The norm proves to be a convenient tool:

Proposition 4.9. *Let* $\mathfrak{O}$ *be the ring of integers in a number field* K, *and let* $x, y \in \mathfrak{O}$. *Then*

(a) *x is a unit if and only if* $\mathrm{N}(x) = \pm 1$,
(b) *If x and y are associates, then* $\mathrm{N}(x) = \pm \mathrm{N}(y)$,
(c) *If* $\mathrm{N}(x)$ *is a rational prime, then x is irreducible in* $\mathfrak{O}$.

Proof. (a) If $xu = 1$, then $\mathrm{N}(x)\mathrm{N}(u) = 1$. Since $\mathrm{N}(x), \mathrm{N}(u) \in \mathbf{Z}$, we have $\mathrm{N}(x) = \pm 1$. Conversely, if $\mathrm{N}(x) = \pm 1$, then

$$\sigma_1(x)\sigma_2(x) \ldots \sigma_n(x) = \pm 1$$

where the σ_i are the monomorphisms $K \to \mathbf{C}$. One factor, without loss in generality $\sigma_1(x)$, is equal to x; all the other $\sigma_i(x)$ are integers. Put

$$u = \pm \sigma^2(x) \ldots \sigma_n(x).$$

Then $xu = 1$, so $u = x^{-1} \in K$. Hence $u \in K \cap \mathbf{B} = \mathfrak{O}$, and x is a unit.

(b) If x, y are associates, then $x = uy$ for a unit u, so $\mathrm{N}(x) = \mathrm{N}(uy) = \mathrm{N}(u)\mathrm{N}(y) = \pm \mathrm{N}(y)$ by (a).

(c) Let $x = yz$. Then $\mathrm{N}(y)\mathrm{N}(z) = \mathrm{N}(yz) = \mathrm{N}(x) = p$, a rational prime; so one of $\mathrm{N}(y)$ and $\mathrm{N}(z)$ is $\pm p$ and the other is ± 1. By (a), one of y and z is a unit, so x is irreducible. □

We have not asserted converses to parts (b) and (c) because these are generally false, as examples in the next section readily reveal.

4.4 Examples of non-unique factorization into irreducibles

We say that factorization in a domain D is *unique* if, whenever

$$p_1 \ldots p_r = q_1 \ldots q_s$$

where every p_i and q_j is irreducible in D, it follows that

(a) $r = s$,

(b) There is a permutation π of $\{1, \ldots, r\}$ such that p_i and $q_{\pi(i)}$ are associates for all $i = 1, \ldots, r$.

In view of our earlier remarks about trivial factorizations, this is the best we can hope for. It says that a factorization into irreducibles (if it exists) is unique except for the order of the factors and the possible presence of units. Variation to this extent is necessary, since even in $\mathbf{Z}$ we have, for instance,

$$3.5 = 5.3 = (-3)(-5) = (-5)(-3).$$

Unfortunately factorization into irreducibles need not be unique in a ring of integers of an algebraic number field. Examples are quite easy to come by if one looks in the right places, and to drive the point home we shall give quite a lot of them. They will be drawn from quadratic fields, and we state them as positive theorems. The easiest come from imaginary quadratic fields:

Theorem 4.10. *Factorization into irreducibles is not unique in the ring of integers of* $\mathbb{Q}(\sqrt{d})$ *for (at least) the following values of* d: $-5, -6, -10, -13, -14, -15, -17, -21, -22, -23, -26, -29, -30$.

Proof. In $\mathbb{Q}(\sqrt{-5})$ we have the factorizations

$$6 = 2.3 = (1 + \sqrt{-5})(1 - \sqrt{-5}).$$

We claim that $2, 3, 1 + \sqrt{-5}$ and $1 - \sqrt{-5}$ are irreducible in the ring $\mathfrak{O}$ of integers of $\mathbb{Q}(\sqrt{-5})$. Since the norm is given by

$$\mathrm{N}(a + b\sqrt{-5}) = a^2 + 5b^2$$

their norms are 4, 9, 6, 6, respectively. If $2 = xy$ where $x, y \in \mathfrak{O}$ are non-units, then $4 = \mathrm{N}(2) = \mathrm{N}(x)\mathrm{N}(y)$ so that $\mathrm{N}(x) = \pm 2$, $\mathrm{N}(y) = \pm 2$. Similarly non-trivial divisors of 3 must, if they exist, have norm ± 3, whilst non-trivial divisors of $1 \pm \sqrt{-5}$ must have norm ± 2 or ± 3. Since $-5 \not\equiv 1 \pmod 4$, the integers in $\mathfrak{O}$ are of the form $a + b\sqrt{-5}$ for $a, b \in \mathbf{Z}$ (Theorem 3.2) so we are led to the equations

$$a^2 + 5b^2 = \pm 2 \text{ or } \pm 3 \qquad (a, b \in \mathbf{Z}).$$

Now $|b| \geqslant 1$ implies $|a^2 + 5b^2| \geqslant 5$, so the only possibility is $|b| = 0$; but then we have $a^2 = \pm 2$ or ± 3, which is impossible in integers. Thus the putative divisors do not exist, and the four factors are all irreducible. Since $\mathrm{N}(2) = 4$, $\mathrm{N}(1 \pm \sqrt{-5}) = 6$, by Proposition 4.9(b) 2 is not an associate of $1 + \sqrt{-5}$ or $1 - \sqrt{-5}$, so factorization is not unique.

The other stated values of d are dealt with in exactly the same way (with a few slight subtleties noted at the end of the proof) starting from the following factorizations:

$$\begin{aligned}
&\mathbf{Q}(\sqrt{-6}): & 6 &= 2.3 &&= (\sqrt{-6})(-\sqrt{-6})\\
&\mathbf{Q}(\sqrt{-10}): & 14 &= 2.7 &&= (2 + \sqrt{-10})(2 - \sqrt{-10})\\
&\mathbf{Q}(\sqrt{-13}): & 14 &= 2.7 &&= (1 + \sqrt{-13})(1 - \sqrt{-13})\\
&\mathbf{Q}(\sqrt{-14}): & 15 &= 3.5 &&= (1 + \sqrt{-14})(1 - \sqrt{-14})\\
&\mathbf{Q}(\sqrt{-15}): & 4 &= 2.2 &&= \left(\frac{1 + \sqrt{-15}}{2}\right)\left(\frac{1 - \sqrt{-15}}{2}\right)\\
&\mathbf{Q}(\sqrt{-17}): & 18 &= 2.3.3 &&= (1 + \sqrt{-17})(1 - \sqrt{-17})\\
&\mathbf{Q}(\sqrt{-21}): & 22 &= 2.11 &&= (1 + \sqrt{-21})(1 - \sqrt{-21})\\
&\mathbf{Q}(\sqrt{-22}): & 26 &= 2.13 &&= (2 + \sqrt{-22})(2 - \sqrt{-22})\\
&\mathbf{Q}(\sqrt{-23}): & 6 &= 2.3 &&= \left(\frac{1 + \sqrt{-23}}{2}\right)\left(\frac{1 - \sqrt{-23}}{2}\right)\\
&\mathbf{Q}(\sqrt{-26}): & 27 &= 3.3.3 &&= (1 + \sqrt{-26})(1 - \sqrt{-26})\\
&\mathbf{Q}(\sqrt{-29}): & 30 &= 2.3.5 &&= (1 + \sqrt{-29})(1 - \sqrt{-29})\\
&\mathbf{Q}(\sqrt{-30}): & 34 &= 2.17 &&= (2 + \sqrt{-30})(2 - \sqrt{-30}).
\end{aligned}$$

Points to note are the following. In cases -15 and -23, note that $d \equiv 1 \pmod 4$ and be careful. For -26 it is easy to prove 3 irreducible. For $1-\sqrt{-26}$ we are led to the equation $N(x)N(y) = 27$, so $N(x) = \pm 9$, $N(y) = \pm 3$, or the other way round. This leads to the equations $a^2 + 26b^2 = \pm 9$ or ± 3. There *is* a solution for ± 9, but not for ± 3, and the latter is sufficient to show $1+\sqrt{-26}$ is irreducible. □

Examining this list, we see that in the ring of integers of $\mathbb{Q}(\sqrt{-17})$ there is an example to show that even the *number* of irreducible factors may differ; the case $\mathbb{Q}(\sqrt{-26})$ shows that the number of distinct factors may differ and that even a (rational) prime power may factorize non-uniquely.

For real quadratic fields there are similar results, but these are harder to find. Also, since the norm is $a^2 - db^2$, it is harder to prove given numbers irreducible. With the same range of values as in Theorem 4.10 we find:

Theorem 4.11. *Factorization into irreducibles is not unique in the ring of integers of* $\mathbb{Q}(\sqrt{d})$ *for (at least) the following values of* d:

$$10, 15, 26, 30.$$

Proof. In the integers of $\mathbb{Q}(\sqrt{10})$ we have factorizations:

$$6 = 2.3 = (4+\sqrt{10})(4-\sqrt{10}).$$

We prove $2, 3, 4 \pm\sqrt{10}$ irreducible. Looking at norms this amounts to proving that the equations

$$a^2 - 10b^2 = \pm 2 \text{ or } \pm 3$$

have no solutions in integers a, b. It is no longer helpful to look at the size of $|b|$, because of the minus sign. However, the equation implies

$$a^2 \equiv \pm 2 \text{ or } \pm 3 \pmod{10}$$

or equivalently

$$a^2 \equiv 2, 3, 7 \text{ or } 8 \pmod{10}.$$

The squares (mod 10) are, in order, 0, 1, 4, 9, 6, 5, 6, 9, 4, 1; by a seemingly remarkable coincidence, the numbers we are looking for are precisely those which do not occur. Hence no solutions exist and the four factors are irreducible. Now 2 and $4 \pm \sqrt{10}$ are not associates, since their norms are 4, 6 respectively.

Similarly we have:

$$\begin{aligned}
\mathbf{Q}(\sqrt{15})&: \ 10 = 2.5 = (5+\sqrt{15})(5-\sqrt{15})\\
\mathbf{Q}(\sqrt{26})&: \ 10 = 2.5 = (6+\sqrt{26})(6-\sqrt{26})\\
\mathbf{Q}(\sqrt{30})&: \ \ 6 = 2.3 = (6+\sqrt{30})(6-\sqrt{30}).
\end{aligned}$$

The reader will find it instructive to do his (or her) own calculations. □

The values of d considered in Theorems 4.10 and 4.11 have not, despite appearances, been chosen at random. If one attempts similar tricks with other d in the range -30 to 30, nothing seems to work. Thus in $\mathbf{Q}(\sqrt{-19})$ we get

$$\left(\frac{1+\sqrt{-19}}{2}\right)\left(\frac{1-\sqrt{-19}}{2}\right) = 5$$

but all this shows is that 5 is reducible.

Trying another obvious product in the integers of $\mathbf{Q}(\sqrt{-19})$, we find

$$(2+\sqrt{-19})(2-\sqrt{-19}) = 23$$

which just tells us that 23 is also reducible. The case

$$\left(\frac{3+\sqrt{-19}}{2}\right)\left(\frac{3-\sqrt{-19}}{2}\right) = 7$$

shows 7 is reducible. After more of these calculations we may alight on

$$35 = 5.7 = (4+\sqrt{-19})(4-\sqrt{-19}).$$

Will this prove non-uniqueness? No, because neither 5 nor 7 is irreducible, as we have seen; and neither is $4 \pm \sqrt{-19}$. The complete factorization of 35 is

$$\left(\frac{1+\sqrt{-19}}{2}\right)\left(\frac{3+\sqrt{-19}}{2}\right)\left(\frac{1-\sqrt{-19}}{2}\right)\left(\frac{3-\sqrt{-19}}{2}\right)$$

and the two apparently distinct factorizations come from different groupings of these in pairs. Eventually one is led to conjecture that the integers of $\mathbf{Q}(\sqrt{-19})$ have unique factorization. This is indeed true, but we shall not be in any position to prove it until Chapter 10. In fact the ring of integers of $\mathbf{Q}(\sqrt{d})$ for negative squarefree d has unique factorization into irreducibles if and only if d takes one of the values:

$$-1, -2, -3, -7, -11, -19, -43, -67, -163.$$

Numerical evidence available in the time of Gauss pointed to this result. In 1934 Heilbronn and Linfoot [29] showed that at most one further value of d could occur, and that $|d|$ had to be very large. In 1952 Heegner [28] offered a proof but it was thought to contain a gap. In 1967 Stark [38] found a proof, as did Baker [16] soon after. And now Birch, Deuring and Siegel [17, 22, 37] have filled in the gap in Heegner's proof. The methods of this book are not appropriate to give any of these proofs, but we will prove in Chapter 10 that for these nine values factorization is unique.

The situation for positive d is not at all well understood. Factorization is unique in many more cases, for instance 2, 3, 5, 6, 7, 11, 13, 14, 17, 19, 21, 22, 23, 29, 31, 33, 37, 38, 41, 43, 46, 47, 53, 57, 59, 61, 62, 67, 69, 71, 73, 77, 83, 86, 89, 93, 94, 97, . . . (these being all for d less than 100). It is not even known whether unique factorization occurs for infinitely many $d > 0$.

So far we have not proved uniqueness of factorization for the ring of integers in any number fields (apart from $\mathbf{Z}$). In the next section we introduce a criterion which tells us when factorization is unique in terms of a special property of the irreducibles.

4.5 Prime factorization

We have already noted that an irreducible p in $\mathbf{Z}$ satisfies the additional property

$$p \mid mn \qquad \text{implies} \qquad p \mid m \text{ or } p \mid n.$$

In this section we shall show that it is this property which characterizes uniqueness of factorization.

In a domain D an element x is said to be *prime* if it is not zero or a unit and

$$x \mid ab \text{ in } D \qquad \text{implies} \qquad x \mid a \text{ or } x \mid b.$$

Note that the zero element satisfies the given property in a domain, but we exclude it to correspond with the definition of prime in $\mathbf{Z}$, where 0 is not usually considered a prime. This convention allows us to state:

Proposition 4.12. *A prime in a domain D is always irreducible.*

Proof. Suppose that D is a domain, $x \in D$ is prime, and $x = ab$. Then $x \mid ab$, so $x \mid a$ or $x \mid b$.

If $x \mid a$, then $a = xc$ ($c \in D$), so

$$x = xcb$$

and cancelling x (which is non-zero), we see that

$$1 = cb$$

and b is a unit. In the same way, $x \mid b$ implies a is a unit. □

The converse of this result is not true, as Eisenstein lamented in 1844; in many domains there exist irreducibles which are not primes. For example in $\mathbf{Z}[\sqrt{-5}]$ we have

$$6 = 2.3 = (1+\sqrt{-5})(1-\sqrt{-5}),$$

but 2 does not divide either of $1+\sqrt{-5}$ or $1-\sqrt{-5}$ (as we saw in the proof of Theorem 4.10). So 2 is an irreducible in $\mathbf{Z}[\sqrt{-5}]$, but not prime. The factorizations in the proofs of Theorems 4.10 and 4.11 readily yield other examples. The next theorem tells us that such examples are entirely typical – every domain with non-unique factorization contains irreducibles which are not prime:

Theorem 4.13. *In a domain in which factorization into irreducibles is possible, factorization is unique if and only if every irreducible is prime.*

Proof. Let D be the domain. It is convenient to rephrase the possibility of factorization for all non-zero $x \in D$ as

$$x = up_1 \dots p_r$$

where u is a unit and $p_1, \dots, p_r$ are irreducibles. When $r = 0$ this can then be interpreted as $x = u$ is a unit and when $r \geqslant 1$, then up_1 is an irreducible, so x is a product of the irreducibles $up_1, p_2, \dots, p_r$.

Now for the proof. Suppose first that factorization is unique and p is an irreducible. We must show p is prime.

$$\text{If } p \mid ab, \text{ then } pc = ab \ (c \in D).$$

We need only consider the non-trivial case $a \neq 0, b \neq 0$ which implies $c \neq 0$ also.

Factorize a, b, c into irreducibles:

$$a = u_1p_1 \dots p_n$$
$$b = u_2q_1 \dots q_m$$
$$c = u_3r_1 \dots r_s$$

where each u_i is a unit and p_i, q_i and r_i are irreducible. Then

$$p(u_3r_1 \dots r_s) = (u_1p_1 \dots p_n)(u_2q_1 \dots q_m),$$

and unique factorization implies p is an associate (hence divides) one of the p_i or q_j, so divides a or b. Hence p is prime.

Conversely, suppose that every irreducible is prime. We will demonstrate that if

$$u_1p_1 \dots p_m = u_2q_1 \dots q_n \tag{2}$$

where u_1, u_2 are units and the p_i, q_j are irreducibles, then $m = n$ and there is a permutation π of $\{1, \ldots, m\}$ such that p_i and $q_{\pi(i)}$ are associates $(1 \leqslant i \leqslant m)$.

This is trivially true for $m = 0$.

For $m \geqslant 1$, if (2) holds, then $p_m \mid u_2 q_1 \ldots q_n$. But p_m is prime, so (by induction on n), $p_m \mid u_2$ or $p_m \mid q_j$ for some j. The first of these possibilities would imply p_m is a unit (by 4.3a), so we must conclude that $p_m \mid q_j$. We renumber so that $j = n$, then $p_m \mid q_n$ and $q_n = p_m u$ where u is a unit. So

$$u_1 p_1 \ldots p_m = u_2 q_1 \ldots q_{n-1} u p_m$$

and cancelling p_m,

$$u_1 p_1 \ldots p_{m-1} = (u_2 u) q_1 \ldots q_{n-1}.$$

By induction we may suppose $m - 1 = n - 1$ and there is a permutation of $1, \ldots, m - 1$ such that p_i, $q_{\pi(i)}$ are associates $(1 \leqslant i \leqslant m - 1)$. We can then extend π to $\{1, \ldots, m\}$ by defining $\pi(m) = m$ to give the required result. □

A domain D is called a *unique factorization domain* if factorization into irreducibles is possible and unique. In a unique factorization domain all irreducibles are primes, so we may speak of a factorization into irreducibles as a 'prime factorization'. Theorem 4.13 tells us that a prime factorization is unique in the usual sense.

We can immediately generalize many ideas on factorization to a unique factorization domain. For instance, if $a, b \in D$, then the highest common factor h of a, b is defined to be an element which satisfies

(i) $h \mid a, h \mid b$,
(ii) If $h' \mid a, h' \mid b$, then $h' \mid h$.

If a is zero, the highest common factor of a, b is b. For $a, b \neq 0$, in a unique factorization domain we can write

$$a = u_1 p_1^{e_1} \ldots p_n^{e_n}$$
$$b = u_2 p_1^{f_1} \ldots p_n^{f_n}$$

where u_1, u_2 are units and the p_i are distinct (i.e.

non-associate) primes with non-negative integer exponents e_i, f_i. Then it is easy to show that

$$h = up_1^{m_1} \dots p_n^{m_n}$$

where u is any unit and m_i is the smaller of e_i, f_i $(1 \leqslant i \leqslant n)$. The highest common factor is unique up to multiplication by a unit. We can say that a, b are *coprime* if their highest common factor is 1 (or any other unit).

In the same way we can define the *lowest common multiple* l of a, b to satisfy

(iii) $a \mid l, b \mid l$,

(iv) If $a \mid l', b \mid l'$, then $l \mid l'$.

For non-zero a, b this is

$$l = up_1^{k_1} \dots p_n^{k_n}$$

where k_i is the larger of e_i, f_i in the factorizations noted above.

Without uniqueness of factorization we can no longer guarantee the existence of highest common factors and lowest common multiples. (See Exercise 4.9.) The language of factorization of integers can only be carried over sensibly to a unique factorization domain.

In the next section we shall see that if a domain has a property analogous to the division algorithm, then we can show that every irreducible is prime, so factorization is unique. In later chapters we shall develop more advanced techniques which will show unique factorization in a wider class of domains which do not have this property.

4.6 Euclidean domains

The crucial property in the usual proofs of unique factorization in $\mathbf{Z}$ or $K[t]$ for a field K is the existence of a division algorithm. A reasonable generalization of this is the following:

Let D be a domain. A *Euclidean function* for D is a function $\phi : D \backslash \{0\} \rightarrow \mathbf{N}$ such that

(i) If $a, b \in D \backslash \{0\}$ and $a \mid b$ then $\phi(a) \leqslant \phi(b)$,

(ii) If $a, b \in D \backslash \{0\}$ then there exist $q, r \in \mathbf{D}$ such that $a = bq + r$ where either $r = 0$ or $\phi(r) < \phi(b)$.

Thus for $\mathbf{Z}$ the function $\phi(n) = |n|$ and for $K[t]$, $\phi(p) = \partial p$ (the degree of the polynomial p) are Euclidean functions.

If a domain has a Euclidean function we call it a *Euclidean domain.* We shall prove that a Euclidean domain has unique factorization by showing that every irreducible is prime. The route is this: first we show that in a Euclidean domain every ideal is principal (a domain with this property is called a *principal ideal domain*), then we show that the latter property implies all irreducibles are primes.

Theorem 4.14. *Every Euclidean domain is a principal ideal domain.*

Proof. Let D be Euclidean, I an ideal of D. If $I = 0$ it is principal, so we may assume there exists a non-zero element x of I. Further choose x to make $\phi(x)$ as small as possible. If $y \in I$ then by (ii) we have $y = qx + r$ where either $r = 0$ or $\phi(r) < \phi(x)$. Now $r \in I$ so we cannot have $\phi(r) < \phi(x)$ because $\phi(x)$ is minimal. This means that $r = 0$, so y is a multiple of x. Therefore $I = \langle x \rangle$ is principal. □

Theorem 4.15. *Every principal ideal domain is a unique factorization domain.*

Proof. Let D be a principal ideal domain. Since this implies D is noetherian, factorization into irreducibles is possible by Theorem 4.6. To prove uniqueness we show that every irreducible is prime.

Suppose p is irreducible, then $\langle p \rangle$ is maximal amongst the principal ideals of D by 4.4(d), but since every ideal is principal, this means that $\langle p \rangle$ is maximal amongst all ideals.

Suppose $p \mid ab$ but $p \nmid a$. The fact that $p \nmid a$ implies $\langle p, a \rangle \supsetneqq \langle p \rangle$, so by maximality, $\langle p, a \rangle = D$.

Then $1 \in \langle p, a \rangle$, so

$$1 = cp + da \ (c, d \in D).$$

Multiplying by b yields

$$b = cpb + dab$$

and since $p \mid ab$, we find $p \mid (cpb + dab)$, so $p \mid b$. This shows p to be prime and completes the proof. □

Theorem 4.16. *A Euclidean domain is a unique factorization domain.* □

4.7 Euclidean quadratic fields

This subsection may be omitted if desired.

In order to apply Theorem 4.16 it is necessary to exhibit some number fields for which the ring of integers is Euclidean. We restrict ourselves to the simplest case of quadratic fields $\mathbf{Q}(\sqrt{d})$ for squarefree d, beginning with the easier situation when d is negative.

Theorem 4.17. *The ring of integers* $\mathfrak{O}$ *of* $\mathbf{Q}(\sqrt{d})$ *is Euclidean for* $d = -1, -2, -3, -7, -11$, *with Euclidean function*

$$\phi(\alpha) = |\mathrm{N}(\alpha)|.$$

Proof. To begin with we consider the suitability of the function ϕ defined in the theorem. For this to be a Euclidean function, the following two conditions must be satisfied for all $\alpha, \beta \in \mathfrak{O}\backslash 0$:

(a) If $\alpha \mid \beta$ then $|\mathrm{N}(\alpha)| \leqslant |\mathrm{N}(\beta)|$.

(b) There exist $\gamma, \delta \in \mathfrak{O}$ such that $\alpha = \beta\gamma + \delta$ where either $\delta = 0$ or $|\mathrm{N}(\delta)| < |\mathrm{N}(\beta)|$.

It is clear that (a) holds, for if $\alpha \mid \beta$ then $\beta = \lambda\alpha$ for $\lambda \in \mathfrak{O}$ and then

$$|\mathrm{N}(\beta)| = |\mathrm{N}(\alpha\lambda)| = |\mathrm{N}(\alpha)\mathrm{N}(\lambda)| = |\mathrm{N}(\alpha)||\mathrm{N}(\lambda)|$$

with rational integer values for the $|\mathrm{N}(?)|$'s.

To prove (b), we consider the alternative statement:

(c) For any $\epsilon \in \mathbf{Q}(\sqrt{d})$ there exists $\kappa \in \mathfrak{O}$ such that

$$|\mathrm{N}(\epsilon - \kappa)| < 1.$$

We shall prove (c) is equivalent to (b). First suppose (b) holds. We know (Lemma 2.10) that $c\epsilon \in \mathfrak{O}$ for some $c \in \mathbf{Z}$. Applying (b) with $\alpha = c\epsilon$, $\beta = c$ we get two possibilities:

(i) $\delta = 0$ and $c\epsilon = c\gamma$ for $\gamma \in \mathfrak{O}$. Then $\epsilon = \gamma \in \mathfrak{O}$ and we may take $\kappa = \epsilon$.

(ii) $c\epsilon = c\gamma + \delta$ where $|\mathrm{N}(\delta)| < |\mathrm{N}(c)|$. Now $c \neq 0$, so this implies

$$|\mathrm{N}(\delta/c)| < 1$$

which is the same as

$$|\mathrm{N}(\epsilon - \gamma)| < 1$$

so we may take $\kappa = \gamma$. Hence (b) implies (c). To prove that (c) implies (b) we put $\epsilon = \alpha/\beta$ and argue similarly.

This allows us to concentrate on condition (c), which is relatively easy to handle: in spirit it says that everything in $\mathbf{Q}(\sqrt{d})$ is 'near to' an integer.

Suppose $\epsilon = r + s\sqrt{d}$ $(r, s \in \mathbf{Q})$. If $d \not\equiv 1 \pmod 4$ we have to find $\kappa = x + y\sqrt{d}$ $(x, y \in \mathbf{Z})$ with

$$|(r-x)^2 - d(s-y)^2| < 1.$$

For $d = -1, -2$ we may do this by taking x and y to be the rational integers nearest to r and s respectively, for then

$$|(r-x)^2 - d(s-y)^2| \leqslant |(\tfrac{1}{2})^2 + 2(\tfrac{1}{2})^2| = \tfrac{3}{4} < 1.$$

The remaining three values of d to be considered have $d \equiv 1 \pmod 4$. In this case we must find

$$\kappa = x + y\left(\frac{1+\sqrt{d}}{2}\right) \qquad (x, y \in \mathbf{Z})$$

such that

$$|(r - x - \tfrac{1}{2}y)^2 - d(s - \tfrac{1}{2}y)^2| < 1.$$

Certainly we can take y to be the rational integer nearest to $2s$, so that $|2s - y| \leqslant \frac{1}{2}$; and then we may find $x \in \mathbf{Z}$ so that $|r - x - \frac{1}{2}y| \leqslant \frac{1}{2}$. For $d = -3, -7$, or -11 this means that

$$|(r - x - \tfrac{1}{2}y)^2 - d(s - \tfrac{1}{2}y)^2| \leqslant |\tfrac{1}{4} + \tfrac{11}{16}| = \tfrac{15}{16} < 1.$$

The theorem is proved. □

To complete the picture for negative d we have:

Theorem 4.18. *For squarefree $d < -11$ the ring of integers of $\mathbf{Q}(\sqrt{d})$ is not Euclidean.*

Proof. Let $\mathfrak{D}$ be the ring of integers of $\mathbf{Q}(\sqrt{d})$ and suppose for a contradiction that there exists a Euclidean function ϕ. (We do *not* assume $\phi = |\mathrm{N}|$.) Choose $\alpha \in \mathfrak{D}$ such that $\alpha \neq 0$, α is not a unit, and $\phi(\alpha)$ is minimal subject to this. Let β be any element of $\mathfrak{D}$. Now there exist γ, δ such that $\beta = \alpha\gamma + \delta$ with $\delta = 0$ or $\phi(\delta) < \phi(\alpha)$. By choice of α the latter condition implies that either $\delta = 0$ or δ is a unit.

For $d < -11$ it is easy to see that the only units of $\mathbf{Q}(\sqrt{d})$ are ± 1 (Proposition 4.2). Hence for every $\beta \in \mathfrak{D}$ we have $\beta \equiv -1, 0$, or $1 \pmod{\langle\alpha\rangle}$ and so $|\mathfrak{D}/\langle\alpha\rangle| \leqslant 3$.

Now we compute $|\mathfrak{D}/\langle\alpha\rangle|$ using Theorem 1.13. By Theorem 2.15 $(\mathfrak{D}, +)$ is free abelian of rank 2. If $d \not\equiv 1 \pmod 4$ a $\mathbf{Z}$-basis for $\langle\alpha\rangle$ is $\{\alpha, \alpha\sqrt{d}\}$ since a $\mathbf{Z}$-basis for $\mathfrak{D}$ is $\{1, \sqrt{d}\}$. If $\alpha = a + b\sqrt{d}$ $(a, b \in \mathbf{Z})$ the $\mathbf{Z}$-basis for $\langle\alpha\rangle$ is

$$\{a + b\sqrt{d},\ db + a\sqrt{d}\}.$$

Hence by Theorem 1.13 we have

$$|\mathfrak{D}/\langle\alpha\rangle| = \left\| \begin{matrix} a & b \\ db & a \end{matrix} \right\| = |a^2 - db^2| = |\mathrm{N}(\alpha)|.$$

Similar calculations apply for $d \equiv 1 \pmod 4$ with the same end result. (These calculations are a special case of Corollary 5.9). It follows that $|\mathrm{N}(\alpha)| \leqslant 3$. Thus if $d \not\equiv 1 \pmod 4$ we have $|a^2 - db^2| \leqslant 3$ with $a, b \in \mathbf{Z}$. If $d \equiv 1 \pmod 4$ then $a = A/2$, $b = B/2$ for $A, B \in \mathbf{Z}$; and then $|A^2 - dB^2| \leqslant 12$. Since $d < -11$ the only solutions are $a = \pm 1$, $b = 0$; so $|\mathrm{N}(\alpha)| = 1$ and hence α is a unit. This contradicts the choice of α. □

These two theorems together show that for negative d the ring of integers of $\mathbf{Q}(\sqrt{d})$ is Euclidean if and only if $d = -1$, $-2, -3, -7, -11$. Further, when it is Euclidean it has as Euclidean function the absolute value of the norm. For brevity call such fields *norm-Euclidean*.

The determination of the norm-Euclidean quadratic fields

with d positive has been a long process involving many mathematicians. Dickson proved $\mathbf{Q}(\sqrt{d})$ Euclidean for $d = 2, 3, 5, 13$ (mistakenly asserting there were no others); Perron added 6, 7, 11, 17, 21, 29 to the list; Oppenheimer, Remak, and Rédei added 19, 33, 37, 41, 55, 73. Rédei claimed 97 as well but this was disproved by Barnes and Swinnerton-Dyer. Heilbronn proved the list finite in 1934, and the problem was finished off by Chatland and Davenport [21] in 1950 (and Inkeri [30] in 1949, independently) who proved:

Theorem 4.19. *The ring of integers of* $\mathbf{Q}(\sqrt{d})$, *for positive* d, *is norm-Euclidean if and only if* $d = 2, 3, 5, 6, 7, 11, 13, 17, 19, 21, 29, 33, 37, 41, 55, 73$. □

We cannot prove this theorem here. A good survey of the problem and related questions, with references, is given by Narkiewicz [33].

Unlike the case d negative it is not known whether $\mathbf{Q}(\sqrt{d})$ can be Euclidean but not norm-Euclidean. In an interesting and readable paper Samuel [35] suggests that $\mathbf{Q}(\sqrt{14})$ might have such properties.

4.8 Consequences of unique factorization

When the integers in a number field have unique factorization, we can carry over many arguments of the type used in the factorization of integers (taking a little care at first). For example, if the reader refers back to the proof of the statement of Fermat on page 74, he will see that, since $\mathbf{Z}[\sqrt{-2}]$ (the ring of integers in $\mathbf{Q}(\sqrt{-2})$) has unique factorization, then the intuitive 'proof' given there is, in fact, valid. Let us consider another example of the same sort of thing, again a statement of Fermat, which we prove:

Theorem 4.20. *The equation*

$$y^2 + 4 = z^3 \tag{9}$$

has only the integer solutions $y = \pm 11, z = 5$, *or* $y = \pm 2$, $z = 2$.

Proof. First suppose y odd, and work in the ring $\mathbf{Z}[i]$, which is a unique factorization domain (Theorem 4.17). Then (9) factorizes as

$$(2 + iy)(2 - iy) = z^3.$$

A common factor $a + ib$ of $2 + iy$, $2 - iy$ is also a factor of their sum, 4, and difference, $2y$, so taking norms

$$a^2 + b^2 \mid 16, \qquad a^2 + b^2 \mid 4y^2,$$

implying

$$a^2 + b^2 \mid 4.$$

The only solutions of this relation are $a = \pm 1, b = 0$, or $a = 0, b = \pm 1$, or $a = \pm 1, b = \pm 1$, none of which turn out to give a proper factor $a + ib$ of $2 + iy$. Hence $2 + iy$, $2 - iy$ are coprime. By unique factorization in $\mathbf{Z}[i]$ it follows that if their product is a cube then one is $\epsilon\alpha^3$ and the other $\epsilon^{-1}\beta^3$ where ϵ is a unit, and $\alpha, \beta \in \mathbf{Z}[i]$. But the units in $\mathbf{Z}[i]$ are $\pm i, \pm 1$ (Proposition 4.2), which are all cubes, so

$$2 + iy = (a + ib)^3$$

for some $a, b \in \mathbf{Z}$. Taking complex conjugates, we find that

$$2 - iy = (a - ib)^3.$$

Adding the two equations,

$$4 = 2a(a^2 - 3b^2)$$

so that

$$a(a^2 - 3b^2) = 2.$$

Now a divides 2, so $a = \pm 1$ or ± 2; and the choice of a determines b. It is easy to see that the only solutions are $a = -1$, $b = \pm 1$, or $a = 2, b = \pm 1$. Then

$$z^3 = ((a + ib)(a - ib))^3 = (a^2 + b^2)^3,$$

so $z = a^2 + b^2 = 2, 5$ respectively. Then $y^2 + 4 = 8, 125$, so $y = \pm 2, \pm 11$. This gives the solutions with $y = \pm 11$ as the only ones for y odd.

Now suppose y even, so that $y = 2Y$. Then z is even as well, say $z = 2Z$, and

$$Y^2 + 1 = 2Z^3.$$

Then Y must be odd, say $Y = 2k + 1$. The highest common factor of $Y + i$ and $Y - i$ divides the difference $2i = (1 + i)^2$. Now $1 + i$ divides $Y + i$ and $Y - i$ but $(1 + i)^2$ does not, so the highest common factor of $Y + i$ and $Y - i$ is $1 + i$. Now

$$(1 + iY)(1 - iY) = 2Z^3$$

and the common factor $1 + i$ occurs twice on the left (bearing in mind that $1 + iY = i(Y - i)$, $1 - iY = -i(Y + i)$). Hence there must be a factorization

$$1 + iY = (1 + i)(a + ib)^3$$

whence as before

$$1 = (a + b)(a^2 - 4ab + b^2)$$

so $a = \pm 1, b = 0$, or $a = 0, b = \pm 1$. These imply $y = \pm 2$, which correspond to the other two solutions stated. □

4.9 The Ramanujan–Nagell theorem

We now give a more intricate and impressive example of how unique factorization properties of algebraic number fields are used to prove theorems on Diophantine equations. Using the uniqueness of factorization in $\mathbf{Q}(\sqrt{-7})$ Nagell verified a conjecture of Ramanujan:

Theorem 4.21. *The only solutions of the equation*

$$x^2 + 7 = 2^n$$

in integers x, n are:

$$\begin{array}{rcccccc} \pm x & = & 1 & 3 & 5 & 11 & 181 \\ n & = & 3 & 4 & 5 & 7 & 15. \end{array}$$

Proof. We work in $\mathbf{Q}(\sqrt{-7})$, whose ring of integers has unique factorization by Theorem 4.17. Clearly a solution for x is odd and we will suppose x is positive.

Assume first that n is even: then we have a factorization of integers:

$$(2^{n/2}+x)(2^{n/2}-x) = 7$$

so that

$$2^{n/2}+x = 7, \qquad 2^{n/2}-x = 1,$$

so

$$2^{1+n/2} = 8$$

and $n = 4, x = 3$.

Now let n be odd, and assume $n > 3$. We have the factorization into primes:

$$2 = \left(\frac{1+\sqrt{-7}}{2}\right)\left(\frac{1-\sqrt{-7}}{2}\right).$$

Now x is odd, $x = 2k+1$, so $x^2+7 = 4k^2+4k+8$ is divisible by 4. Putting $m = n-2$, we can rewrite the equation to be solved as

$$\frac{x^2+7}{4} = 2^m$$

so that

$$\left(\frac{x+\sqrt{-7}}{2}\right)\left(\frac{x-\sqrt{-7}}{2}\right) = \left(\frac{1+\sqrt{-7}}{2}\right)^m\left(\frac{1-\sqrt{-7}}{2}\right)^m$$

where the right-hand side is a prime factorization. Neither $(1+\sqrt{-7})/2$ nor $(1-\sqrt{-7})/2$ is a common factor of the terms on the left because such a factor would divide their difference, $\sqrt{-7}$, which is seen to be impossible by taking norms. Comparing the two factorizations, since the only units in the integers of $\mathbf{Q}(\sqrt{-7})$ are ± 1, (Proposition 4.2), we must have

$$\frac{x\pm\sqrt{-7}}{2} = \pm\left(\frac{1\pm\sqrt{-7}}{2}\right)^m$$

from which we derive

$$\pm\sqrt{-7} = \left(\frac{1+\sqrt{-7}}{2}\right)^m - \left(\frac{1-\sqrt{-7}}{2}\right)^m.$$

We claim that the positive sign cannot occur. For, putting $\left(\frac{1+\sqrt{-7}}{2}\right) = a$, $\left(\frac{1-\sqrt{-7}}{2}\right) = b$, we have

$$a^m - b^m = a - b.$$

Then

$$a^2 \equiv (1-b)^2 \equiv 1 \pmod{b^2}$$

since $ab = 2$, and so

$$a^m \equiv a(a^2)^{(m-1)/2} \equiv a \pmod{b^2}$$

whence

$$a \equiv a - b \pmod{b^2},$$

a contradiction.

Hence the sign must be negative, and expanding by the binomial theorem we have

$$-2^{m-1} = \binom{m}{1} - \binom{m}{3}7 + \binom{m}{5}7^2 \dots \pm \binom{m}{m}7^{(m-1)/2}$$

so that

$$-2^{m-1} \equiv m \pmod{7}. \qquad (10)$$

Now $2^6 \equiv 1 \pmod{7}$ and it follows easily that the only solutions of (10) are

$$m \equiv 3, 5, \text{ or } 13 \pmod{42}.$$

We prove that only $m \equiv 3, 5, 13$ can occur. It suffices to show that we cannot have two solutions of the original equation which are congruent modulo 42. So let m, m_1 be two such solutions, and let 7^l be the largest power of 7 dividing $m - m_1$. Then

$$a^{m_1} = a^m a^{m_1 - m} = a^m (\tfrac{1}{2})^{m_1 - m}(1 + \sqrt{-7})^{m_1 - m}. \qquad (11)$$

Now

$$(\tfrac{1}{2})^{m_1 - m} = [(\tfrac{1}{2})^6]^{(m_1 - m)/6} \equiv 1 \pmod{7^{l+1}},$$

and

$$(1 + \sqrt{-7})^{m_1 - m} \equiv 1 + (m_1 - m)\sqrt{-7} \pmod{7^{l+1}}$$

(first raise to powers $7, 7^2, \dots, 7^l$, then $(m - m_1)/7^l$). Since

$$a^m \equiv \frac{1 + m\sqrt{-7}}{2^m} \pmod{7}$$

substituting in (11) gives

$$a^{m_1} \equiv a^m + \frac{m_1 - m}{2^m}\sqrt{-7} \pmod{7^{l+1}}$$

and

$$b^{m_1} \equiv b^m - \frac{m_1 - m}{2^m}\sqrt{-7} \qquad (\text{mod } 7^{l+1}).$$

But

$$a^m - b^m = a^{m_1} - b^{m_1}$$

so

$$(m - m_1)\sqrt{-7} \equiv 0 \qquad (\text{mod } 7^{l+1}).$$

Since m and m_1 are rational integers,

$$m \equiv m_1 \qquad (\text{mod } 7^{l+1})$$

which contradicts the definition of l. □

Exercises

4.1 Which of the following elements of $\mathbf{Z}[i]$ are irreducible ($i = \sqrt{-1}$): $1 + i, 3 - 7i, 5, 7, 12i, -4 + 5i$?

4.2 Write down the group of units of the ring of integers of: $\mathbf{Q}(\sqrt{-1})$, $\mathbf{Q}(\sqrt{-2})$, $\mathbf{Q}(\sqrt{-3})$, $\mathbf{Q}(\sqrt{-5})$, $\mathbf{Q}(\sqrt{-6})$.

4.3 Is the group of units of the integers in $\mathbf{Q}(\sqrt{3})$ finite?

4.4 Show that a homomorphic image of a noetherian ring is noetherian.

4.5 Find all ideals of $\mathbf{Z}$ which contain $\langle 120 \rangle$. Show that every ascending chain of ideals of $\mathbf{Z}$ starting with $\langle 120 \rangle$ stops, by direct examination of the possibilities.

4.6 Find a ring which is not noetherian.

4.7 Check the calculations required to complete Theorems 4.10, 4.11.

4.8 Is $10 = (3 + i)(3 - i) = 2.5$ an example of non-unique factorization in $\mathbf{Z}[i]$? Give reasons for your answer.

4.9 Show that 6 and $2(1+\sqrt{-5})$ both have 2 and $1+\sqrt{-5}$ as factors, but do not have a highest common factor in $\mathbf{Z}[\sqrt{-5}]$. Do they have a least common multiple? (consider norms.)

4.10 Let D be any integral domain. Suppose an element $x \in D$ has a factorization

$$x = up_1 \ldots p_n$$

where u is a unit and $p_1, \ldots, p_n$ are *primes*. Show that given any factorization

$$x = vq_1 \ldots q_m$$

where v is a unit and $q_1, \ldots, q_m$ are *irreducibles,* then $m = n$ and there exists a permutation π of $\{1, \ldots, n\}$ such that $p_i, q_{\pi(i)}$ are associates ($1 \leqslant i \leqslant n$).

4.11 Show in $\mathbf{Z}[\sqrt{-5}]$ that $\sqrt{-5} \mid (a + b\sqrt{-5})$ if and only if $5 \mid a$. Deduce that $\sqrt{-5}$ is prime in $\mathbf{Z}[\sqrt{-5}]$. Hence conclude that the element 5 factorizes uniquely into irreducibles in $\mathbf{Z}[\sqrt{-5}]$ although $\mathbf{Z}[\sqrt{-5}]$ does not have unique factorization.

4.12 Suppose D is a unique factorization domain, and a, b are coprime non-units. Deduce that if

$$ab = c^n$$

for $c \in D$, then there exists a unit $e \in D$ such that ea and $e^{-1}b$ are nth powers in D.

4.13 Let p be an odd rational prime and $\zeta = e^{2\pi i/p}$. If α is a prime element in $\mathbf{Z}[\zeta]$, prove that the rational integers which are divisible by α are precisely the rational integer multiples of some prime rational integer q. (Hint: $\alpha \mid N(\alpha)$, so α divides some rational prime factor q of $N(\alpha)$. Now show α is not a factor of any $m \in \mathbf{Z}$ prime to q.)

4.14 Prove that the ring of integers of $\mathbf{Q}(e^{2\pi i/5})$ is Euclidean.

4.15 Prove that the ring of integers of $\mathbf{Q}(\sqrt{2}, i)$ is Euclidean.

4.16 Let $\mathbf{Q}_2$ be the set of all rational numbers a/b, where $a, b \in \mathbf{Z}$ and b is odd. Prove that $\mathbf{Q}_2$ is a domain, and that the only irreducibles in $\mathbf{Q}_2$ are 2 and its associates.

4.17 Generalize 4.16 to the ring $\mathbf{Q}_\pi$, where π is a finite set of ordinary primes, this being defined as the set of all rationals a/b with b prime to the elements of π.

4.18 The following purports to be a proof that in any number field K the ring of integers contains infinitely many irreducibles. Find the error.
'Assume $\mathfrak{O}$ has only finitely many irreducibles $p_1, \ldots, p_n$. The number $1 + p_1 \ldots p_n$ must be divisible by some irreducible q, and this cannot be any of $p_1, \ldots, p_n$. This is a contradiction. Of course the argument breaks down unless we can find at least one irreducible in $\mathfrak{O}$; but since not every element of $\mathfrak{O}$ is a unit this is easy: let x be any non-unit and let p be some irreducible factor of x.'
Hint: The 'proof' does not use any properties of $\mathfrak{O}$ beyond the existence of irreducible factorization and the fact that not every element is a unit. Now $\mathbf{Q}_2$ has these properties . . .

4.19 Give a correct proof of the statement in Exercise 4.18.

CHAPTER FIVE

Ideals

Following the trail blazed by Kummer and Dedekind, we show that although uniqueness of factorization may fail to hold for elements in $\mathfrak{O}$, a unique factorization theory works for ideals. In this theory the essential building blocks are prime ideals, which are defined by adapting the definition of prime element from the previous chapter. We also generalize the concept of an ideal slightly to that of 'fractional ideal'. The advantage of this approach is that the non-zero fractional ideals form a group under multiplication; from this the uniqueness of factorization into prime ideals follows easily. The multiplicative group of non-zero fractional ideals, or more accurately one of its quotient groups, will play an increasingly important part in the theory.

Several standard consequences of unique factorization are easily deduced. Next we define the norm of an ideal, as a generalization of the norm of an element of $\mathfrak{O}$, and prove that the new norm has a multiplicative property. We also prove that every ideal of $\mathfrak{O}$ can be generated by at most two elements. This allows us to show that factorization of elements of $\mathfrak{O}$ into irreducibles is unique if and only if every ideal is principal, thereby linking the two factorization theories: ideals and elements.

5.1 Historical background

To motivate Kummer's introduction of 'ideal numbers' and Dedekind's reformulation of this concept in terms of 'ideals', we shall look more closely at some examples of the failure of unique factorization, in the hope that some pattern may emerge.

Many of our previous examples exhibit no obvious pattern, but others do seem to have significant features. For instance, consider:

$$\mathbb{Q}(\sqrt{15}): \qquad 2.5 = (5+\sqrt{15})(5-\sqrt{15})$$
$$\mathbb{Q}(\sqrt{30}): \qquad 2.3 = (6+\sqrt{30})(6-\sqrt{30})$$
$$\mathbb{Q}(\sqrt{-10}): \qquad 2.7 = (2+\sqrt{-10})(2-\sqrt{-10}).$$

In these we see a curious phenomenon: there is a prime p occurring on the left, and on the right a factor $a + b\sqrt{d}$ where a and d are multiples of p. One feels somehow that $\sqrt{p}$ is a common factor of both sides, but $\sqrt{p}$ does not lie in the given number field. As a specific case, consider the first example where $\sqrt{5}$ looks a likely candidate for a common factor, but $\sqrt{5}$ is not an element of $\mathbb{Q}(\sqrt{15})$. Leaving aside the niceties at the moment, introduce $\sqrt{5}$ into the factorization to get

$$5+\sqrt{15} = \sqrt{5}(\sqrt{5}+\sqrt{3})$$
$$5-\sqrt{15} = \sqrt{5}(\sqrt{5}-\sqrt{3}).$$

Multiplying up and cancelling the 5, we get

$$2 = (\sqrt{5}+\sqrt{3})(\sqrt{5}-\sqrt{3}).$$

We can now see that the two given factorizations of 10 are obtained by grouping the factors in

$$(\sqrt{5})(\sqrt{5})(\sqrt{5}+\sqrt{3})(\sqrt{5}-\sqrt{3})$$

in two different ways.

Perhaps by introducing new numbers, such as $\sqrt{5}$, we can restore unique factorization. Can our problem be that we are not factorizing in the right context? In other words, if factorization of some element in the ring of integers of the given number field K is not unique, can we extend K to a field L where it *is*? In our example to factorize the element 10 we extended $\mathbb{Q}(\sqrt{15})$ to $\mathbb{Q}(\sqrt{3},\sqrt{5})$. What about the others? The factorizations of 14 in $\mathbb{Q}(\sqrt{-10})$ can be found by extending to $\mathbb{Q}(\sqrt{2},\sqrt{-5})$ to get two possible groupings of the factors in

$$14 = (\sqrt{2})(\sqrt{2})(\sqrt{2}+\sqrt{-5})(\sqrt{2}-\sqrt{-5}).$$

The case of 6 in $\mathbb{Q}(\sqrt{30})$ is even more interesting: we have

$$6 = (\sqrt{2})(\sqrt{2})(\sqrt{3})(\sqrt{3})(\sqrt{6}+\sqrt{5})(\sqrt{6}-\sqrt{5}),$$

and the last two factors are *units*.

This is one way of viewing Kummer's theory. One starts with a number field K and extends to a field L. Then $\mathfrak{O}_K \subseteq \mathfrak{O}_L$. Neither of these rings of integers need have unique factorization, but an element in $\mathfrak{O}_K$ may factorize uniquely into elements in $\mathfrak{O}_L$.

At the outset Kummer did not describe the theory in this way. His method involved detailed computations, which are described in [2]. This involved him introducing the notion of 'ideal' prime factors for elements which may have no actual prime factors in $\mathfrak{O}_K$ at all. These additional 'ideal' numbers may be interpreted as the elements introduced from $\mathfrak{O}_L$ for factorization purposes.

A more appropriate formulation of the theory by Dedekind in terms of ideals (in the modern sense, though the origin of the name goes back to Kummer's theory) has clarified matters. To motivate this approach, consider a factorization of an element

$$x = ab$$

in a ring R. Recalling from Chapter 1 that the product of ideals IJ is just the set of finite sums $\Sigma x_i y_i$ ($x_i \in I$, $y_i \in J$), we see that the ideal generated by x is the product of the ideals generated by a and by b:

$$\langle x\rangle = \langle a\rangle\langle b\rangle.$$

More generally a product

$$x = p_1 \ldots p_n$$

of elements in R corresponds to a product of principal ideals

$$\langle x \rangle = \langle p_1 \rangle \ldots \langle p_n \rangle.$$

In considering uniqueness of factorization, the formulation in terms of ideals is marginally better, for if we replace p_1 by up_1 where u is a unit, we find that the ideals $\langle p_1 \rangle$ and $\langle up_1 \rangle$ are the same (Proposition 4.4b). Thus if the factors are unique up to multiplication by units and order, the ideals $\langle p_1 \rangle, \ldots, \langle p_n \rangle$ are unique up to order. By passing to ideals we eliminate the problems introduced by units.

How does this tie in with the earlier discussion? First consider the example

$$10 = (\sqrt{5})(\sqrt{5})(\sqrt{5}+\sqrt{3})(\sqrt{5}-\sqrt{3})$$

in the integers of $\mathbf{Q}(\sqrt{3},\sqrt{5})$. Let $K = \mathbf{Q}(\sqrt{15})$, $L = \mathbf{Q}(\sqrt{3}, \sqrt{5})$; then this factorization holds in the ring of integers $\mathfrak{O}_L$. In this ring we also have the corresponding factorization of principal ideals:

$$\langle 10 \rangle = \langle \sqrt{5} \rangle \langle \sqrt{5} \rangle \langle \sqrt{5}+\sqrt{3} \rangle \langle \sqrt{5}-\sqrt{3} \rangle.$$

We may intersect the ideals in this factorization with $\mathfrak{O}_K$, and once more we get ideals in $\mathfrak{O}_K$, but now these ideals may not be principal. For instance, let $I = \langle \sqrt{5}+\sqrt{3} \rangle \cap \mathfrak{O}_K$. Then $\sqrt{3}(\sqrt{5}+\sqrt{3}) = \sqrt{15}+3 \in I$, and $\sqrt{5}(\sqrt{5}+\sqrt{3}) = 5+\sqrt{15} \in I$, so their difference

$$(5+\sqrt{15})-(3-\sqrt{15}) = 2 \in I.$$

If I were principal, say $I = \langle a+b\sqrt{15} \rangle$, then 2 would be a multiple of $a+b\sqrt{15}$, and taking norms,

$$a^2 - 15b^2 \mid 4.$$

Suppose that I is principal, say $I = \langle k \rangle$. Now $N(5+\sqrt{15}) = 10$ and $N(3+\sqrt{15}) = -6$, so $N(k) \mid 2$. We know that $N(k) \neq \pm 1$ since I is proper. If $N(k) = \pm 2$ then there exist $a, b \in \mathbf{Z}$ with $a^2 - 15b^2 = \pm 2$. But, taken mod 5, this leads to a contradiction. So I is not principal.

The moral is now clear. If we wish to factorize the principal ideal $\langle x\rangle$ in a ring of integers $\mathfrak{O}_K$, then we may get a unique factorization of ideals,

$$\langle x\rangle = I_1 \dots I_n,$$

but the ideals $I_1, \dots, I_n$ may not be principal.

Factorization into ideals proves to be most useful, however; for the ideals in $\mathfrak{O}_K$ are not far off being principal, having (as we shall see) at most two generators.

5.2 Prime factorization of ideals

Throughout this chapter $\mathfrak{O}$ is the ring of integers of a number field K of degree n. We use small Gothic letters to denote ideals (and later 'fractional ideals') of $\mathfrak{O}$. We are interested in two special types of ideal, which we define in a general situation as follows. Let R be a ring. Then an ideal $\mathfrak{a}$ of R is *maximal* if $\mathfrak{a}$ is a proper ideal of R and there are no ideals of R strictly between $\mathfrak{a}$ and R. The ideal $\mathfrak{a} \neq R$ of R is *prime* if, whenever $\mathfrak{b}$ and $\mathfrak{c}$ are ideals of R with $\mathfrak{bc} \subseteq \mathfrak{a}$, then either $\mathfrak{b} \subseteq \mathfrak{a}$ or $\mathfrak{c} \subseteq \mathfrak{a}$.

We can see where the latter definition comes from by considering the special case where all three ideals concerned are principal, say $\mathfrak{a} = \langle a\rangle$, $\mathfrak{b} = \langle b\rangle$, $\mathfrak{c} = \langle c\rangle$. Since $x \mid y$ is equivalent to $\langle y\rangle \subseteq \langle x\rangle$ (Proposition 4.4a), then the statement

$$\mathfrak{bc} \subseteq \mathfrak{a} \text{ implies either } \mathfrak{b} \subseteq \mathfrak{a} \text{ or } \mathfrak{c} \subseteq \mathfrak{a}$$

translates into

$$a \mid bc \text{ implies either } a \mid b \text{ or } a \mid c.$$

If R is an integral domain, then the zero ideal is prime, and here we find $\langle p\rangle$ is prime if and only if p is a prime or zero. (See Exercise 5.1.) The fact that we exclude 0 from the list of prime elements but include $\langle 0\rangle$ as a prime ideal is a quirk of the historical development of the subject. Elements came first and 0 was excluded from the list of primes of $\mathbf{Z}$. On the other hand, the definition we have given for a prime ideal allows us to make the following simple characterizations:

Lemma 5.1. *Let R be a ring and $\mathfrak{a}$ an ideal of R. Then*
(a) *$\mathfrak{a}$ is maximal if and only if $R/\mathfrak{a}$ is a field,*
(b) *$\mathfrak{a}$ is prime if and only if $R/\mathfrak{a}$ is a domain.*

Proof. The ideals of $R/\mathfrak{a}$ are in bijective correspondence with the ideals of R lying between $\mathfrak{a}$ and R. Hence $\mathfrak{a}$ is maximal if and only if $R/\mathfrak{a}$ has no non-zero proper ideals. Now it is easy to show that a ring S has no non-zero proper ideals if and only if S is a field. Taking $S = R/\mathfrak{a}$ proves (a).

To prove (b), first suppose $\mathfrak{a}$ is prime. If $x, y \in R$ are such that in $R/\mathfrak{a}$ we have

$$(\mathfrak{a} + x)(\mathfrak{a}+y) = 0$$

and then $xy \in \mathfrak{a}$, so $\langle x\rangle\langle y\rangle \subseteq \mathfrak{a}$. Hence either $\langle x\rangle \subseteq \mathfrak{a}$ or $\langle y\rangle \subseteq \mathfrak{a}$, so either $x \in \mathfrak{a}$ or $y \in \mathfrak{a}$. Hence one of $(\mathfrak{a} + x)$ or $(\mathfrak{a} + y)$ is zero in $R/\mathfrak{a}$, and therefore $R/\mathfrak{a}$ has no zero-divisors so is a domain. Conversely suppose $R/\mathfrak{a}$ is a domain. Then $|R/\mathfrak{a}| \neq 1$ so $\mathfrak{a} \neq R$. Suppose if possible that $\mathfrak{b}\mathfrak{c} \subseteq \mathfrak{a}$ but $\mathfrak{b} \nsubseteq \mathfrak{a}, \mathfrak{c} \nsubseteq \mathfrak{a}$. Then we can find elements $b \in \mathfrak{b}$, $c \in \mathfrak{c}$, with $b, c \notin \mathfrak{a}$ but $bc \in \mathfrak{a}$. This means that $(\mathfrak{a} + b)$ and $(\mathfrak{a} + c)$ are zero-divisors in $R/\mathfrak{a}$, which is a contradiction. □

Corollary 5.2. *Every maximal ideal is prime.* □

Next we list some important properties of the ring of integers of a number field:

Theorem 5.3. *The ring of integers $\mathfrak{O}$ of a number field K has the following properties:*
(a) *It is a domain, with field of fractions K,*
(b) *It is noetherian,*
(c) *If $\alpha \in K$ satisfies a monic polynomial equation with coefficients in $\mathfrak{O}$ then $\alpha \in \mathfrak{O}$,*
(d) *Every non-zero prime ideal of $\mathfrak{O}$ is maximal.*

Proof. Part (a) is obvious. For part (b) note that by Theorem 2.15 the group $(\mathfrak{O}, +)$ is free abelian of rank n. It follows by Theorem 1.12 that if $\mathfrak{a}$ is an ideal of $\mathfrak{O}$ then $(\mathfrak{a}, +)$ is free abelian of rank $\leqslant n$. Now any $\mathbf{Z}$-basis for $(\mathfrak{a}, +)$ generates $\mathfrak{a}$ as an ideal, so every ideal of $\mathfrak{O}$ is finitely generated and $\mathfrak{O}$ is noetherian. Part (c) is immediate from Theorem 2.9. To prove part (d) let $\mathfrak{p}$ be a prime ideal of $\mathfrak{O}$. Let $0 \neq \alpha \in \mathfrak{p}$. Then

$$N = \mathrm{N}(\alpha) = \alpha_1 \ldots \alpha_n \in \mathfrak{p}$$

(the α_i being the conjugates of α) since $\alpha_1 = \alpha$. Therefore $\langle N \rangle \subseteq \mathfrak{p}$, and hence $\mathfrak{O}/\mathfrak{p}$ is a quotient ring of $\mathfrak{O}/N\mathfrak{O}$ which, being a finitely generated abelian group with every element of finite order, is finite. Since $\mathfrak{O}/\mathfrak{p}$ is a domain by Lemma 5.1(b) and is finite, it is a field by Theorem 1.1. Hence $\mathfrak{p}$ is a maximal ideal by Lemma 5.1(a). □

The reader should note that 5.3(d) is by no means typical of general rings. For example if we take $R = \mathbf{R}[x, y]$, the ring of polynomials in indeterminates x, y with real coefficients, then the ideal $\langle x \rangle$ is prime but not maximal because $R/\langle x \rangle \cong \mathbf{R}[y]$ is a domain which is not a field. A ring which satisfies the conditions 5.3(a)-(d) is called a *Dedekind ring* after the mathematician who made ring-theoretic advances in this area. The proof of unique factorization of ideals which we shall give shortly will hold good in all Dedekind rings, although in applications we shall only require the special case when the ring is a ring $\mathfrak{O}$ of integers in a number field.

To prove uniqueness we need to study the 'arithmetic' of non-zero ideals of $\mathfrak{O}$, especially their behaviour under multiplication. Clearly this multiplication is commutative and associative with $\mathfrak{O}$ itself as an identity. However, inverses need not exist, so we do not have a group structure. It turns out that we can capture a group if we spread our net wider. Note that an ideal may be described as an $\mathfrak{O}$-submodule of $\mathfrak{O}$, so we look at $\mathfrak{O}$-submodules of the field K. The particular submodules of interest to give the group structure we desire will turn out to be characterized by the following property: an $\mathfrak{O}$-submodule $\mathfrak{a}$ of K is called a *fractional ideal* of $\mathfrak{O}$ if there exists some non-zero $c \in \mathfrak{O}$ such that $c\mathfrak{a} \subseteq \mathfrak{O}$. In other words, the set $\mathfrak{b} = c\mathfrak{a}$ is an ideal of $\mathfrak{O}$, and $\mathfrak{a} = c^{-1}\mathfrak{b}$; thus the fractional ideals of $\mathfrak{O}$ are subsets of K of the form $c^{-1}\mathfrak{b}$

where $\mathfrak{b}$ is an ideal of $\mathfrak{O}$ and c is a non-zero element of $\mathfrak{O}$. (This explains the name.)

Example. The fractional ideals of $\mathbf{Z}$ are of the form $r\mathbf{Z}$ where $r \in \mathbf{Q}$.

Of course if every ideal of $\mathfrak{O}$ is principal, then the fractional ideals are of the form $c^{-1}\langle d\rangle = c^{-1}d\,\mathfrak{O}$ where d is a generator. By 5.3(a) this means the fractional ideals in a principal ideal domain $\mathfrak{O}$ are just $\alpha\,\mathfrak{O}$ where $\alpha \in K$. The interest in fractional ideals is greater because $\mathfrak{O}$ need not be a principal ideal domain.

In general, an ideal is clearly a fractional ideal and, conversely, a fractional ideal $\mathfrak{a}$ is an ideal if and only if $\mathfrak{a} \subseteq \mathfrak{O}$. The product of fractional ideals is once more a fractional ideal. In fact, if $\mathfrak{a}_1 = c_1^{-1}\,\mathfrak{b}_1, \mathfrak{a}_2 = c_2^{-1}\,\mathfrak{b}_2$ where $\mathfrak{b}_1, \mathfrak{b}_2$ are ideals and c_1, c_2 are non-zero elements of $\mathfrak{O}$, then $\mathfrak{a}_1\mathfrak{a}_2 = (c_1c_2)^{-1}\,\mathfrak{b}_1\mathfrak{b}_2$. The multiplication of fractional ideals is commutative and associative with $\mathfrak{O}$ acting as an identity.

Theorem 5.4. *The non-zero fractional ideals of $\mathfrak{O}$ form an abelian group under multiplication.*

It is convenient to prove this result along with the main theorem of the chapter:

Theorem 5.5. *Every non-zero ideal of $\mathfrak{O}$ can be written as a product of prime ideals, uniquely up to the order of the factors.*

Proof. We shall prove 5.4 and 5.5 together in a series of steps.

(i) *Let $\mathfrak{a} \neq 0$ be an ideal of $\mathfrak{O}$. Then there exist prime ideals $\mathfrak{p}_1, \ldots, \mathfrak{p}_r$ such that $\mathfrak{p}_1 \ldots \mathfrak{p}_r \subseteq \mathfrak{a}$.*

For a contradiction suppose not. Then since $\mathfrak{O}$ is noetherian (Theorem 5.3b) we may choose $\mathfrak{a}$ maximal, subject to the non-existence of such $\mathfrak{p}$'s. Then $\mathfrak{a}$ is not prime (since we could then take $\mathfrak{p}_1 = \mathfrak{a}$), so there exist ideals $\mathfrak{b}, \mathfrak{c}$ of $\mathfrak{O}$ with $\mathfrak{b}\mathfrak{c} \subseteq \mathfrak{a}$, $\mathfrak{b} \nsubseteq \mathfrak{a}$, $\mathfrak{c} \nsubseteq \mathfrak{a}$. Let

$$\mathfrak{a}_1 = \mathfrak{a} + \mathfrak{b}, \qquad \mathfrak{a}_2 = \mathfrak{a} + \mathfrak{c}.$$

Then $\mathfrak{a}_1\mathfrak{a}_2 \subseteq \mathfrak{a}$, $\mathfrak{a}_1 \supsetneqq \mathfrak{a}$, $\mathfrak{a}_2 \supsetneqq \mathfrak{a}$. By maximality of $\mathfrak{a}$ there exist prime ideals $\mathfrak{p}_1, \ldots, \mathfrak{p}_s, \mathfrak{p}_{s+1}, \ldots, \mathfrak{p}_r$ such that

$$\mathfrak{p}_1 \ldots \mathfrak{p}_s \subseteq \mathfrak{a}_1,$$

$$\mathfrak{p}_{s+1} \ldots \mathfrak{p}_r \subseteq \mathfrak{a}_2.$$

Hence

$$\mathfrak{p}_1 \ldots \mathfrak{p}_r \subseteq \mathfrak{a}_1\mathfrak{a}_2 \subseteq \mathfrak{a}$$

contrary to the choice of $\mathfrak{a}$.

(ii) *Definition of what will turn out to be the inverse of an ideal:*

For each ideal $\mathfrak{a}$ of $\mathfrak{O}$, define

$$\mathfrak{a}^{-1} = \{x \in K \mid x\mathfrak{a} \subseteq \mathfrak{O}\}$$

It is clear that $\mathfrak{a}^{-1}$ is an $\mathfrak{O}$-submodule. If $\mathfrak{a} \neq 0$ then for any $c \in \mathfrak{a}$, $c \neq 0$, we have $c\mathfrak{a}^{-1} \subseteq \mathfrak{O}$, so $\mathfrak{a}^{-1}$ is a fractional ideal. Clearly $\mathfrak{O} \subseteq \mathfrak{a}^{-1}$, so $\mathfrak{a} = \mathfrak{a}\mathfrak{O} \subseteq \mathfrak{a}\mathfrak{a}^{-1}$. From the definition we have

$$\mathfrak{a}\mathfrak{a}^{-1} = \mathfrak{a}^{-1}\mathfrak{a} \subseteq \mathfrak{O}.$$

This means that the fractional ideal $\mathfrak{a}\mathfrak{a}^{-1}$ is actually an *ideal*. (Our aim will be to prove $\mathfrak{a}\mathfrak{a}^{-1} = \mathfrak{O}$.) A further useful fact for ideals $\mathfrak{p}$, $\mathfrak{a}$ is that $\mathfrak{a} \subseteq \mathfrak{p}$ implies $\mathfrak{O} \subseteq \mathfrak{p}^{-1} \subseteq \mathfrak{a}^{-1}$.

(iii) *If $\mathfrak{a}$ is a proper ideal, then $\mathfrak{a}^{-1} \supsetneqq \mathfrak{O}$.*

Since $\mathfrak{a} \subseteq \mathfrak{p}$ for some maximal ideal $\mathfrak{p}$, whence $\mathfrak{p}^{-1} \subseteq \mathfrak{a}^{-1}$, it is sufficient to prove $\mathfrak{p}^{-1} \neq \mathfrak{O}$ for $\mathfrak{p}$ maximal. We must therefore find a non-integer in $\mathfrak{p}^{-1}$. We start with any $a \in \mathfrak{p}$, $a \neq 0$. Using (i) we choose the smallest r such that

$$\mathfrak{p}_1 \ldots \mathfrak{p}_r \subseteq \langle a \rangle$$

for $\mathfrak{p}_1, \ldots, \mathfrak{p}_r$ prime. Since $\langle a \rangle \subseteq \mathfrak{p}$ and $\mathfrak{p}$ is prime (remember maximal implies prime), some $\mathfrak{p}_i \subseteq \mathfrak{p}$. Without loss of generality $\mathfrak{p}_1 \subseteq \mathfrak{p}$. Hence $\mathfrak{p}_1 = \mathfrak{p}$ since prime ideals in $\mathfrak{O}$ are maximal (Theorem 5.3d) and further

$$\mathfrak{p}_2 \ldots \mathfrak{p}_r \nsubseteq \langle a \rangle$$

by minimality of r. Hence we can find $b \in \mathfrak{p}_2 \ldots \mathfrak{p}_r \backslash \langle a \rangle$. But

$b\mathfrak{p} \subseteq \langle a \rangle$ so $ba^{-1}\mathfrak{p} \subseteq \mathfrak{D}$ and $ba^{-1} \in \mathfrak{p}^{-1}$. But $b \notin a\,\mathfrak{D}$ and so $ba^{-1} \notin \mathfrak{D}$, whence $\mathfrak{p}^{-1} \neq \mathfrak{D}$.

(iv) *If* $\mathfrak{a}$ *is a non-zero ideal and* $\mathfrak{a}S \subseteq \mathfrak{a}$ *for any subset* $S \subseteq K$, *then* $S \subseteq \mathfrak{D}$.

We must show that if $\mathfrak{a}\theta \subseteq \mathfrak{a}$ for $\theta \in S$, then $\theta \in \mathfrak{D}$. Because $\mathfrak{D}$ is noetherian, $\mathfrak{a} = \langle a_1, \ldots, a_m \rangle$, where not all the a_i are zero. Then $\mathfrak{a}\theta \subseteq \mathfrak{a}$ implies

$$\begin{aligned} a_1\theta &= b_{11}a_1 + \ldots + b_{1m}a_m \\ &\;\;\ldots\ldots \qquad\qquad\qquad\qquad (b_{ij} \in \mathfrak{D}) \\ a_m\theta &= b_{m1}a_1 + \ldots + b_{mm}a_m. \end{aligned}$$

As in Lemma 2.7 we deduce that because the equations

$$\begin{aligned} (b_{11} - \theta)x_1 + \ldots + b_{1m}x_m &= 0 \\ \ldots\ldots \\ b_{m1}x_1 + \ldots + (b_{mm} - \theta)x_m &= 0 \end{aligned}$$

have a non-zero solution $x_1 = a_1, \ldots, x_m = a_m$, then the determinant of the array of coefficients is non-zero. This gives a monic polynomial equation in θ with coefficients in $\mathfrak{D}$, hence $\theta \in \mathfrak{D}$ by Theorem 5.3(c). (We remark that we could short-cut part of this proof by noting, as in the proof of Theorem 5.3, that the b_{ij} may be taken to be *rational* integers which gives $\theta \in \mathfrak{D}$ directly.)

We are now in a position to take an important step in the proof of Theorem 5.4:

(v) *If* $\mathfrak{p}$ *is a maximal ideal, then* $\mathfrak{p}\mathfrak{p}^{-1} = \mathfrak{D}$.

From (ii), $\mathfrak{p}\mathfrak{p}^{-1}$ is an ideal where $\mathfrak{p} \subseteq \mathfrak{p}\mathfrak{p}^{-1} \subseteq \mathfrak{D}$. Since $\mathfrak{p}$ is maximal, $\mathfrak{p}\mathfrak{p}^{-1}$ is equal to $\mathfrak{p}$ or $\mathfrak{D}$. But if $\mathfrak{p}\mathfrak{p}^{-1} = \mathfrak{p}$, then (iv) would imply $\mathfrak{p}^{-1} \subseteq \mathfrak{D}$, contradicting (iii). So $\mathfrak{p}\mathfrak{p}^{-1} = \mathfrak{D}$.

We can now extend (iv) to any ideal $\mathfrak{a}$:

(vi) *For every ideal* $\mathfrak{a} \neq 0$, $\mathfrak{a}\mathfrak{a}^{-1} = \mathfrak{D}$.

If not, choose $\mathfrak{a}$ maximal subject to $\mathfrak{a}\mathfrak{a}^{-1} \neq \mathfrak{D}$. Then $\mathfrak{a} \subseteq \mathfrak{p}$ where $\mathfrak{p}$ is maximal. From (ii), $\mathfrak{D} \subseteq \mathfrak{p}^{-1} \subseteq \mathfrak{a}^{-1}$, so

$$\mathfrak{a} \subseteq \mathfrak{a}\mathfrak{p}^{-1} \subseteq \mathfrak{a}\mathfrak{a}^{-1} \subseteq \mathfrak{D}.$$

In particular, $\mathfrak{a}\mathfrak{p}^{-1} \subseteq \mathfrak{D}$ implies $\mathfrak{a}\mathfrak{p}^{-1}$ is an ideal. Now we cannot have $\mathfrak{a} = \mathfrak{a}\mathfrak{p}^{-1}$, for that would imply $\mathfrak{p}^{-1} \subseteq \mathfrak{D}$ by (iv), contradicting (iii) once more. So $\mathfrak{a} \subsetneqq \mathfrak{a}\mathfrak{p}^{-1}$ and the maximality condition on $\mathfrak{a}$ implies the ideal $\mathfrak{a}\mathfrak{p}^{-1}$ satisfies

$$\mathfrak{a}\mathfrak{p}^{-1}(\mathfrak{a}\mathfrak{p}^{-1})^{-1} = \mathfrak{O}.$$

By the definition of $\mathfrak{a}^{-1}$ this means

$$\mathfrak{p}^{-1}(\mathfrak{a}\mathfrak{p}^{-1})^{-1} \subseteq \mathfrak{a}^{-1}.$$

Thus

$$\mathfrak{O} = \mathfrak{a}\mathfrak{p}^{-1}(\mathfrak{a}\mathfrak{p}^{-1})^{-1} \subseteq \mathfrak{a}\mathfrak{a}^{-1} \subseteq \mathfrak{O}$$

from which the result follows.

(vii) *Every fractional ideal* $\mathfrak{a}$ *has an inverse* $\mathfrak{a}^{-1}$ *such that* $\mathfrak{a}\mathfrak{a}^{-1} = \mathfrak{O}$.

The set $\mathscr{F}$ of fractional ideals is already known to be a commutative semigroup, so given a fractional ideal $\mathfrak{a}$, we only need to find another fractional ideal $\mathfrak{a}'$ such that $\mathfrak{a}\mathfrak{a}' = \mathfrak{O}$, then $\mathfrak{a}'$ will be the required inverse. But there exists an ideal $\mathfrak{b}$ and a non-zero element $c \in \mathfrak{O}$ such that $\mathfrak{a} = c^{-1}\mathfrak{b}$. Let $\mathfrak{a}' = c\mathfrak{b}^{-1}$, then $\mathfrak{a}\mathfrak{a}' = \mathfrak{O}$ as required.

This, of course, proves Theorem 5.4.

(viii) *Every non-zero ideal* $\mathfrak{a}$ *is a product of prime ideals.*

If not, let $\mathfrak{a}$ be maximal subject to the condition of not being a product of prime ideals. Then $\mathfrak{a}$ is not prime, but we will have $\mathfrak{a} \subseteq \mathfrak{p}$ for some maximal (hence prime) ideal, and as in (vi),

$$\mathfrak{a} \subsetneqq \mathfrak{a}\mathfrak{p}^{-1} \subseteq \mathfrak{O}.$$

By the maximality condition on $\mathfrak{a}$,

$$\mathfrak{a}\mathfrak{p}^{-1} = \mathfrak{p}_2 \ldots \mathfrak{p}_r$$

for prime ideals $\mathfrak{p}_2, \ldots, \mathfrak{p}_r$, whence

$$\mathfrak{a} = \mathfrak{p}\mathfrak{p}_2 \ldots \mathfrak{p}_r.$$

(ix) *Prime factorization is unique.*

By analogy with factorization of elements, for ideals $\mathfrak{a}$, $\mathfrak{b}$ we shall say that $\mathfrak{a}$ divides $\mathfrak{b}$ (written $\mathfrak{a} \mid \mathfrak{b}$) if there is an ideal $\mathfrak{c}$ such that $\mathfrak{b} = \mathfrak{a}\mathfrak{c}$. This condition is equivalent to $\mathfrak{a} \supseteq \mathfrak{b}$ since we may then take $\mathfrak{c} = \mathfrak{a}^{-1}\mathfrak{b}$. The definition of prime ideal $\mathfrak{p}$ shows that if $\mathfrak{p} \mid \mathfrak{a}\mathfrak{b}$ then either $\mathfrak{p} \mid \mathfrak{a}$ or $\mathfrak{p} \mid \mathfrak{b}$. If we now have prime ideals $\mathfrak{p}_1, \ldots, \mathfrak{p}_r, \mathfrak{q}_1, \ldots, \mathfrak{q}_s$ with

$$\mathfrak{p}_1 \dots \mathfrak{p}_r = \mathfrak{q}_1 \dots \mathfrak{q}_s,$$

then $\mathfrak{p}_1$ divides some $\mathfrak{q}_i$, so by maximality $\mathfrak{p}_1 = \mathfrak{q}_i$. Multiplying by $\mathfrak{p}_1^{-1}$ and using induction we obtain uniqueness of prime factorization up to the order of the factors.

This proves Theorem 5.5. □

In fact, the fractional ideals also factorize uniquely if we allow negative powers of prime ideals. Namely, if $\mathfrak{a}$ is a fractional ideal with $0 \neq c \in \mathfrak{O}$ such that $c\mathfrak{a}$ is an ideal, we have

$$\langle c \rangle = \mathfrak{p}_1 \dots \mathfrak{p}_r, \qquad c\mathfrak{a} = \mathfrak{q}_1 \dots \mathfrak{q}_s,$$

and hence

$$\mathfrak{a} = \mathfrak{p}_1^{-1} \dots \mathfrak{p}_r^{-1} \mathfrak{q}_1 \dots \mathfrak{q}_s.$$

One result in the proofs of 5.4 and 5.5 which is worth isolating occurs in step (ix):

Proposition 5.6. *For ideals* $\mathfrak{a}, \mathfrak{b}$ *of* $\mathfrak{O}$,

$$\mathfrak{a} | \mathfrak{b} \textit{ if and only if } \mathfrak{a} \supseteq \mathfrak{b}.$$ □

This tells us that in $\mathfrak{O}$ the factors of an ideal $\mathfrak{b}$ are precisely the ideals containing $\mathfrak{b}$. The definition of a prime ideal $\mathfrak{p}$ also translates into a notation directly analogous to that of a prime element:

$$\mathfrak{p} | \mathfrak{a}\mathfrak{b} \text{ implies } \mathfrak{p} | \mathfrak{a} \text{ or } \mathfrak{p} | \mathfrak{b}.$$

An extended worked example. The factorization of the ideal $\langle 18 \rangle$ in $\mathbf{Z}[\sqrt{-17}]$.

From Theorem 4.10 we have the factorization of elements:

$$18 = 2.3.3. = (1 + \sqrt{-17})(1 - \sqrt{-17}).$$

Consider the ideal $\mathfrak{p}_1 = \langle 2, 1 + \sqrt{-17} \rangle$ whose generators are both factors of 18. Clearly $18 \in \mathfrak{p}_1$, so $\langle 18 \rangle \subseteq \mathfrak{p}_1$, which means that $\mathfrak{p}_1$ is a factor of $\langle 18 \rangle$. In fact we also have

$$1 - \sqrt{-17} = 2 - (1 + \sqrt{-17}) \in \mathfrak{p}_1$$

so

$$18 = (1 + \sqrt{-17})(1 - \sqrt{-17}) \in \mathfrak{p}_1^2$$

which means that $\langle 18 \rangle \subseteq \mathfrak{p}_1^2$ and $\mathfrak{p}_1^2$ is a factor of $\langle 18 \rangle$. Now the elements of $\mathfrak{p}_1$ are of the form

$$\begin{aligned} 2(a+b\sqrt{-17}) + (1+\sqrt{-17})(c+d\sqrt{-17}) \\ = (2a+c-17d)+(2b+c+d)\sqrt{-17} \\ = r+s\sqrt{-17} \end{aligned}$$

where

$$r-s = 2a-2b-18d,$$

which is always even. Clearly r may be taken to be any integer and then s may be any integer of the same parity (odd or even). This implies $\mathfrak{p}_1$ is not the whole ring $\mathbf{Z}[\sqrt{-17}]$. On the other hand, $\mathfrak{p}_1$ is maximal, for if $m+n\sqrt{-17}$ is any element not in $\mathfrak{p}_1$ then one of m, n is even and the other odd, so

$$\langle \mathfrak{p}_1, m+n\sqrt{-17}\rangle = \mathbf{Z}[\sqrt{-17}].$$

Similarly, considering

$$\mathfrak{p}_2 = \langle 3, 1+\sqrt{-17}\rangle,$$

we find that an element of $\mathfrak{p}_2$ is of the form

$$r+s\sqrt{-17} = (3a+c-17d)+(3b+c+d)\sqrt{-17}$$

where $r-s=3(a+b-6d)$. Thus r, s can be any integers subject to the constraint

$$r \equiv s \pmod 3.$$

Once more we find $\mathfrak{p}_2$ maximal and $18 = 2.3.3 \in \mathfrak{p}_2^2$, so $\mathfrak{p}_2^2$ is a factor of $\langle 18\rangle$.

Finally, considering

$$\mathfrak{p}_3 = \langle 3, 1-\sqrt{-17}\rangle,$$

we get another prime ideal such that $\mathfrak{p}_3^2$ is a factor of $\langle 18\rangle$, and a calculation similar to the previous ones shows that $r+s\sqrt{-17} \in \mathfrak{p}_3$ if and only if

$$r+s \equiv 0 \pmod 3.$$

Using the factorization theory of Theorem 5.5, we find that

$$\mathfrak{p}_1^2\mathfrak{p}_2^2\mathfrak{p}_3^2 \supseteq \langle 18\rangle.$$

The final step, to show that $\langle 18\rangle = \mathfrak{p}_1^2\mathfrak{p}_2^2\mathfrak{p}_3^2$, is best performed using a counting argument. Since every element in $\mathbf{Z}[\sqrt{-17}]$ is either in $\mathfrak{p}_1$ or of the form $1 + x$ for $x \in \mathfrak{p}_1$, the number of elements in the quotient ring $\mathbf{Z}[-17]/\mathfrak{p}_1$ is

$$|\mathbf{Z}[\sqrt{-17}]/\mathfrak{p}_1| = 2.$$

Similarly

$$|\mathbf{Z}[\sqrt{-17}]/\mathfrak{p}_r| = 3 \quad (r = 2, 3).$$

In the next section we shall call

$$|\mathfrak{O}/\mathfrak{p}|$$

the *norm* of the ideal $\mathfrak{p}$ and write it as $N(\mathfrak{p})$. The crucial property of this new type of norm is that it is multiplicative,

$$N(\mathfrak{a}\mathfrak{b}) = N(\mathfrak{a})N(\mathfrak{b}).$$

Granted this fact, we can deduce

$$N(\mathfrak{p}_1^2\mathfrak{p}_2^2\mathfrak{p}_3^2) = 2^2.3^2.3^2 = 18^2.$$

Now the norm of the ideal $\langle 18\rangle$ is

$$N(\langle 18\rangle) = |\mathbf{Z}[\sqrt{-17}]/\langle 18\rangle|$$

and since every element of $\mathbf{Z}[\sqrt{-17}]$ is uniquely of the form

$$a + b\sqrt{-17} + x$$

where a, b are integers in the range 0 to 17 and $x \in \langle 18\rangle$, we find 18 choices each for a, b so

$$N(\langle 18\rangle) = 18^2.$$

Suppose $\langle 18\rangle$ factorizes as

$$\langle 18\rangle = \mathfrak{p}_1^2\mathfrak{p}_2^2\mathfrak{p}_3^2\mathfrak{a}$$

for some ideal $\mathfrak{a}$. Then taking norms and using the multiplicative property, we find $N(\mathfrak{a}) = 1$, whence $\mathfrak{a}$ is the whole ring and

$$\langle 18\rangle = \mathfrak{p}_1^2\mathfrak{p}_2^2\mathfrak{p}_3^2. \tag{1}$$

If we consider the factorization of elements $18 = 2.3.3$, we obtain

$$\langle 2\rangle\langle 3\rangle^2 = \mathfrak{p}_1^2\mathfrak{p}_2^2\mathfrak{p}_3^2. \tag{2}$$

By the uniqueness of factorization of ideals, both $\langle 2\rangle$, $\langle 3\rangle$ are products of prime ideals from the set $\{\mathfrak{p}_1, \mathfrak{p}_2, \mathfrak{p}_3\}$. Now $2 \in \mathfrak{p}_1$ but $2 \notin \mathfrak{p}_2$, $2 \notin \mathfrak{p}_3$, so $\mathfrak{p}_1 \mid \langle 2\rangle$, $\mathfrak{p}_2 \nmid \langle 2\rangle$, $\mathfrak{p}_3 \nmid \langle 2\rangle$, thus

$$\langle 2\rangle = \mathfrak{p}_1^q.$$

Similarly $3 \notin \mathfrak{p}_1$, $3 \in \mathfrak{p}_2$, $3 \in \mathfrak{p}_3$ implies

$$\langle 3\rangle = \mathfrak{p}_2^r\mathfrak{p}_3^s.$$

Substituting in (2) gives

$$\mathfrak{p}_1^q\mathfrak{p}_2^{2r}\mathfrak{p}_3^{2s} = \mathfrak{p}_1^2\mathfrak{p}_2^2\mathfrak{p}_3^2,$$

and uniqueness of factorization of ideals implies

$$q = 2,\ r = s = 1,$$

$$\langle 2\rangle = \mathfrak{p}_1^2,\ \langle 3\rangle = \mathfrak{p}_2\mathfrak{p}_3. \tag{3}$$

(The reader may find it instructive to check these by direct calculation.)

A similar argument using

$$\langle 18\rangle = \langle 1+\sqrt{-17}\rangle\langle 1-\sqrt{-17}\rangle = \mathfrak{p}_1^2\mathfrak{p}_2^2\mathfrak{p}_3^2 \tag{4}$$

where $1+\sqrt{-17} \in \mathfrak{p}_1, \mathfrak{p}_2$; $1+\sqrt{-17} \notin \mathfrak{p}_3$; $1-\sqrt{-17} \in \mathfrak{p}_1, \mathfrak{p}_3$; $1-\sqrt{-17} \notin \mathfrak{p}_2$ gives

$$\langle 1+\sqrt{-17}\rangle = \mathfrak{p}_1^m\mathfrak{p}_2^n,\ \langle 1-\sqrt{-17}\rangle = \mathfrak{p}_1^r\mathfrak{p}_3^s.$$

Substituting in (4) implies $m = r = 1$, $n = s = 2$, so

$$\langle 1+\sqrt{-17}\rangle = \mathfrak{p}_1\mathfrak{p}_2^2,\ \langle 1-\sqrt{-17}\rangle = \mathfrak{p}_1\mathfrak{p}_3^2. \tag{5}$$

From (3) and (5) we see that the two alternative factorizations of the element 18 come from alternative groupings of the ideals:

$$\begin{aligned}\langle 18\rangle &= (\mathfrak{p}_1^2)(\mathfrak{p}_2\mathfrak{p}_3)^2 = \langle 2\rangle\langle 3\rangle^2\\ &= (\mathfrak{p}_1\mathfrak{p}_2^2)(\mathfrak{p}_1\mathfrak{p}_3^2) = \langle 1+\sqrt{-17}\rangle\langle 1-\sqrt{-17}\rangle.\end{aligned}$$

We shall consider the norm of an ideal and its multiplicative property in the next section, once we have dealt with some simple consequences of unique factorization. Later on we shall develop certain other properties of the norm which will help streamline the calculations in the above example.

5.3 The norm of an ideal

Once unique factorization is proved, several useful consequences follow in the usual way. In particular, any two non-zero ideals $\mathfrak{a}$ and $\mathfrak{b}$ have a *greatest common divisor* $\mathfrak{g}$ and a *least common multiple* $\mathfrak{l}$ with the following properties:

$$\mathfrak{g} \mid \mathfrak{a},\ \mathfrak{g} \mid \mathfrak{b};\ \text{ and if } \mathfrak{g}' \text{ has the same properties } \mathfrak{g}' \mid \mathfrak{g};$$

$$\mathfrak{a} \mid \mathfrak{l},\ \mathfrak{b} \mid \mathfrak{l};\ \text{ and if } \mathfrak{l}' \text{ has the same properties } \mathfrak{l} \mid \mathfrak{l}'.$$

In fact, suppose we factorize $\mathfrak{a}$ and $\mathfrak{b}$ into primes as:

$$\mathfrak{a} = \prod \mathfrak{p}_i^{e_i}, \qquad \mathfrak{b} = \prod \mathfrak{p}_i^{f_i}$$

with distinct prime ideals $\mathfrak{p}_i$. Then we clearly have

$$\mathfrak{g} = \prod \mathfrak{p}_i^{\min(e_i, f_i)}$$

$$\mathfrak{l} = \prod \mathfrak{p}_i^{\max(e_i, f_i)} .$$

We have useful alternative expressions:

Lemma 5.7. *If $\mathfrak{a}$ and $\mathfrak{b}$ are ideals of $\mathfrak{O}$ and $\mathfrak{g}, \mathfrak{l}$ are the greatest common divisor and least common multiple, respectively, of $\mathfrak{a}$ and $\mathfrak{b}$, then*

$$\mathfrak{g} = \mathfrak{a} + \mathfrak{b}, \qquad \mathfrak{l} = \mathfrak{a} \cap \mathfrak{b}.$$

Proof. We know that $\mathfrak{x} \mid \mathfrak{a}$ if and only if $\mathfrak{x} \supseteq \mathfrak{a}$ (Proposition 5.6). Hence $\mathfrak{g}$ must be the smallest ideal containing $\mathfrak{a}$ and $\mathfrak{b}$, and $\mathfrak{c}$ the largest ideal contained in $\mathfrak{a}$ and $\mathfrak{b}$. The rest is obvious. □

The proof of Theorem 5.3 shows that if $\mathfrak{a}$ is a non-zero ideal of $\mathfrak{O}$ then the quotient ring $\mathfrak{O}/\mathfrak{a}$ is finite. We define the *norm* of $\mathfrak{a}$ to be

$$N(\mathfrak{a}) = |\mathfrak{O}/\mathfrak{a}|.$$

Then $N(\mathfrak{a})$ is a positive integer. There is no reason to confuse this norm with the old norm of an element $N(a)$ since it applies only to ideals. In fact there is a connection between the two norms, as we shall see in a moment.

Theorem 5.8. (a) *Every ideal* $\mathfrak{a}$ *of* $\mathfrak{O}$ *with* $\mathfrak{a} \neq 0$ *has a* **Z**-*basis* $\{\alpha_1, \ldots, \alpha_n\}$ *where n is the degree of K,*

(b) *We have*

$$N(\mathfrak{a}) = \left| \frac{\Delta[\alpha_1, \ldots, \alpha_n]}{\Delta} \right|^{1/2}$$

where Δ *is the discriminant of K.*

Proof. We know from Theorem 2.15 that $(\mathfrak{O}, +)$ is free abelian of rank n. Since $\mathfrak{O}/\mathfrak{a}$ is finite it follows from Theorem 1.13 that $(\mathfrak{a}, +)$ is free abelian of rank n, hence has a **Z**-basis of the form $\{\alpha_1, \ldots, \alpha_n\}$. This proves (a). For part (b) let $\{\omega_1, \ldots, \omega_n\}$ be a **Z**-basis for $\mathfrak{O}$, and suppose that $\alpha_i = \Sigma c_{ij}\omega_j$. Then by Theorem 1.13,

$$N(\mathfrak{a}) = |\mathfrak{O}/\mathfrak{a}| = |\det c_{ij}|.$$

But by the formula before Theorem 2.6 we have

$$\begin{aligned} \Delta[\alpha_1, \ldots, \alpha_n] &= (\det c_{ij})^2 \, \Delta[\omega_1, \ldots, \omega_n] \\ &= (N(\mathfrak{a}))^2 \Delta. \end{aligned}$$

Taking square roots and remembering that $N(\mathfrak{a})$ is positive we obtain the desired result. □

Corollary 5.9. *If* $\mathfrak{a} = \langle a \rangle$ *is a principal ideal then* $N(\mathfrak{a}) = |N(a)|$.

Proof. A **Z**-basis for $\mathfrak{a}$ is given by $\{a\omega_1, \ldots, a\omega_n\}$. The result follows from the definition of $\Delta[\alpha_1, \ldots, \alpha_n]$ and the theorem above.

This corollary helps us to make a straightforward calculation of the norm of a principal ideal.

Example. If $\mathfrak{O}$ is the ring of integers of $\mathbf{Q}(\sqrt{d})$ for a square-free rational integer d, then

$$N(\langle a + b\sqrt{d}\rangle) = |a^2 - bd^2|,$$

in particular, in $\mathfrak{O} = \mathbf{Z}[\sqrt{-17}]$, then

$$N(\langle 18\rangle) = 18^2.$$

The new norm, like the old, is *multiplicative*:

Theorem 5.10. *If* $\mathfrak{a}$ *and* $\mathfrak{b}$ *are non-zero ideals of* $\mathfrak{O}$, *then*

$$N(\mathfrak{ab}) = N(\mathfrak{a})N(\mathfrak{b}).$$

Proof. By uniqueness of factorization and induction on the number of factors, it is sufficient to prove

$$N(\mathfrak{ap}) = N(\mathfrak{a})N(\mathfrak{p}) \tag{6}$$

where $\mathfrak{p}$ is prime.

We establish

$$|\mathfrak{O}/\mathfrak{ap}| = |\mathfrak{O}/\mathfrak{a}|\ |\mathfrak{a}/\mathfrak{ap}| \tag{7}$$

and

$$|\mathfrak{a}/\mathfrak{ap}| = |\mathfrak{O}/\mathfrak{p}|. \tag{8}$$

Then (6) follows immediately from (7), (8) and the definition of the norm.

Equation (7) is a consequence of the isomorphism theorem for rings: define $\phi : \mathfrak{O}/\mathfrak{ap} \to \mathfrak{O}/\mathfrak{a}$ by

$$\phi(\mathfrak{ap} + x) = \mathfrak{a} + x$$

then ϕ is a surjective ring homomorphism with kernel $\mathfrak{a}/\mathfrak{ap}$; Lagrange's theorem (applied to the additive groups) gives (7).

To establish (8), first note that unique factorization implies $\mathfrak{a} \neq \mathfrak{ap}$, so $\mathfrak{a} \supsetneqq \mathfrak{ap}$. Now we show that there is no ideal $\mathfrak{b}$ strictly between $\mathfrak{a}$ and $\mathfrak{ap}$, for if

$$\mathfrak{a} \supseteq \mathfrak{b} \supseteq \mathfrak{ap},$$

then, as fractional ideals,

$$\mathfrak{a}^{-1}\mathfrak{a} \supseteq \mathfrak{a}^{-1}\mathfrak{b} \supseteq \mathfrak{a}^{-1}\mathfrak{ap},$$

so

$$\mathfrak{O} \supseteq \mathfrak{a}^{-1}\mathfrak{b} \supseteq \mathfrak{p}.$$

Since $\mathfrak{a}^{-1}\mathfrak{b} \subseteq \mathfrak{O}$, we see that it is actually an ideal, and since $\mathfrak{p}$ is maximal (by 5.3d), we have

$$\mathfrak{a}^{-1}\mathfrak{b} = \mathfrak{O} \text{ or } \mathfrak{a}^{-1}\mathfrak{b} = \mathfrak{p}$$

so

$$\mathfrak{b} = \mathfrak{a} \quad \text{or} \quad \mathfrak{b} = \mathfrak{a}\mathfrak{p}.$$

This means that for any element $a \in \mathfrak{a}\backslash\mathfrak{a}\mathfrak{p}$, we have

$$\mathfrak{a}\mathfrak{p} + \langle a \rangle = \mathfrak{a}. \tag{9}$$

Fix such an a and define $\theta : \mathfrak{O} \to \mathfrak{a}/\mathfrak{a}\mathfrak{p}$ by

$$\theta(x) = \mathfrak{a}\mathfrak{p} + ax,$$

then θ is an $\mathfrak{O}$-module homomorphism, surjective by (9), whose kernel is an ideal satisfying

$$\mathfrak{p} \subseteq \ker \theta.$$

Now $\ker \theta \neq \mathfrak{O}$ (for that would mean $\mathfrak{a}/\mathfrak{a}\mathfrak{p} \cong \mathfrak{O}/\ker\theta = 0$, which would contradict $\mathfrak{a} \neq \mathfrak{a}\mathfrak{p}$), and $\mathfrak{p}$ is maximal, so

$$\ker \theta = \mathfrak{p}.$$

Hence $\mathfrak{O}/\mathfrak{p} \cong \mathfrak{a}/\mathfrak{a}\mathfrak{p}$ (as $\mathfrak{O}$-modules), which gives (8) and completes the proof. □

Example. If $\mathfrak{O} = \mathbf{Z}[\sqrt{-17}]$, $\mathfrak{p}_1 = \langle 2, 1 + \sqrt{-17}\rangle$, $\mathfrak{p}_2 = \langle 3, 1 + \sqrt{-17}\rangle$, $\mathfrak{p}_3 = \langle 3, 1 - \sqrt{-17}\rangle$, then

$$\mathrm{N}(\mathfrak{p}_1^2\mathfrak{p}_2^2\mathfrak{p}_3^2) = 2^2.3^2.3^2 = 18^2.$$

This particular calculation completes the details of the extended example in the previous section.

It is convenient to introduce yet another usage for the word 'divides'. If $\mathfrak{a}$ is an ideal of $\mathfrak{O}$ and b an element of $\mathfrak{O}$ such that $\mathfrak{a} \mid \langle b \rangle$, then we also write $\mathfrak{a} \mid b$ and say that $\mathfrak{a}$ divides b. It is clear that $\mathfrak{a} \mid b$ if and only if $b \in \mathfrak{a}$; however, the new notation has certain distinct advantages. For example, if $\mathfrak{p}$ is a prime ideal and $\mathfrak{p} \mid \langle a \rangle\langle b \rangle$, then we must have $\mathfrak{p} \mid \langle a \rangle$

or $\mathfrak{p} \mid \langle b \rangle$. Thus for $\mathfrak{p}$ prime,

$$\mathfrak{p} \mid ab \text{ implies } \mathfrak{p} \mid a \text{ or } \mathfrak{p} \mid b.$$

This new notation allows us to emphasize the correspondence between factorization of elements and principal ideals which would otherwise be less evident.

Theorem 5.11. *Let* $\mathfrak{a}$ *be an ideal of* $\mathfrak{O}, \mathfrak{a} \neq 0$.

(a) *If* $\mathrm{N}(\mathfrak{a})$ *is prime, then so is* $\mathfrak{a}$.

(b) $\mathrm{N}(\mathfrak{a})$ *is an element of* $\mathfrak{a}$, *or equivalently* $\mathfrak{a} \mid \mathrm{N}(\mathfrak{a})$.

(c) *If* $\mathfrak{a}$ *is prime it divides exactly one rational prime* p, *and then*

$$\mathrm{N}(\mathfrak{a}) = p^m$$

where $m \leqslant n$, *the degree of* K.

Proof. For part (a) write $\mathfrak{a}$ as a product of prime ideals and equate norms. For part (b) note that since $\mathrm{N}(\mathfrak{a}) = |\mathfrak{O}/\mathfrak{a}|$ it follows that for any $x \in \mathfrak{O}$ we have $\mathrm{N}(\mathfrak{a})x \in \mathfrak{a}$. Now put $x = 1$. For part (c) we note that by (b)

$$\mathfrak{a} \mid \mathrm{N}(\mathfrak{a}) = p_1^{m_1} \ldots p_r^{m_r}$$

so (considering principal ideals in place of the p_i) we have $\mathfrak{a} \mid p_i$ for some rational prime p_i. If p and q were distinct rational primes, both divisible by $\mathfrak{a}$, we could find integers u, v such that $up + vq = 1$, and then deduce that $\mathfrak{a} \mid 1$, which implies $\mathfrak{a} = \mathfrak{O}$, a contradiction. Then

$$\mathrm{N}(\mathfrak{a}) \mid \mathrm{N}(\langle p \rangle) = p^n$$

so that $\mathrm{N}(\mathfrak{a}) = p^m$ for some $m \leqslant n$. □

Example 1. If $\mathfrak{O} = \mathbf{Z}[\sqrt{-17}]$, $\mathfrak{p}_1 = \langle 2, 1 + \sqrt{-17} \rangle$, then because $\mathrm{N}(\mathfrak{p}_1) = 2$ we can immediately deduce that $\mathfrak{p}_1$ is prime. Note that $\mathrm{N}(\mathfrak{p}_1) = 2 \in \mathfrak{p}_1$, as asserted by 5.11 (b).

Example 2. A prime ideal $\mathfrak{a}$ can satisfy

$$\mathrm{N}(\mathfrak{a}) = p^m$$

where $m > 1$, which means that a prime ideal does not necessarily have a norm which is prime; for instance $\mathfrak{D} = \mathbf{Z}[i]$, $\mathfrak{a} = \langle 3 \rangle$. Here 3 is irreducible in $\mathbf{Z}[i]$, hence prime because $\mathbf{Z}[i]$ has unique factorization. It is an easy deduction (Exercise 5.1) that if an element is prime, so is the ideal it generates. Hence $\langle 3 \rangle$ is prime in $\mathbf{Z}[i]$, but

$$N(\langle 3 \rangle) = 3^2.$$

The next theorem collects together several useful finiteness assertions:

Theorem 5.12. (a) *Every non-zero ideal of* $\mathfrak{D}$ *has a finite number of divisors,*

(b) *A non-zero rational integer belongs to only a finite number of ideals of* $\mathfrak{D}$*,*

(c) *Only finitely many ideals of* $\mathfrak{D}$ *have given norm.*

Proof. (a) is an immediate consequence of prime factorization, (b) is a special case of (a), and (c) follows from (b) using Theorem 5.11 part (b). □

Example. Consider our earlier calculation

$$\langle 18 \rangle = \mathfrak{p}_1^2 \mathfrak{p}_2^2 \mathfrak{p}_3^2$$

in $\mathbf{Z}[\sqrt{-17}]$ where $\mathfrak{p}_1 = \langle 2, 1+\sqrt{-17} \rangle$, $\mathfrak{p}_2 = \langle 3, 1+\sqrt{-17} \rangle$, $\mathfrak{p}_3 = \langle 3, 1-\sqrt{-17} \rangle$. We find the only prime divisors of $\langle 18 \rangle$ are $\mathfrak{p}_1$, $\mathfrak{p}_2$, $\mathfrak{p}_3$. If 18 belongs to some ideal $\mathfrak{a}$, then $\langle 18 \rangle \subseteq \mathfrak{a}$, whence $\mathfrak{a} \mid \langle 18 \rangle$, so $\mathfrak{a} \mid \mathfrak{p}_1^2 \mathfrak{p}_2^2 \mathfrak{p}_3^2$ and $\mathfrak{a} = \mathfrak{p}_1^q \mathfrak{p}_2^r \mathfrak{p}_3^s$ where q, r, s are 0, 1 or 2. Thus 18 belongs only to a finite number of ideals.

How many ideals $\mathfrak{a}$ have norm 18? This can only happen when $\mathfrak{a} \mid 18$ by 5.11(b), so

$$\mathfrak{a} = \mathfrak{p}_1^q \mathfrak{p}_2^r \mathfrak{p}_3^s$$

which implies

$$N(\mathfrak{a}) = 2^q . 3^r . 3^s.$$

This norm is 18 only when $q = 1$ and $r + s = 2$, which means that $\mathfrak{a}$ is $\mathfrak{p}_1 \mathfrak{p}_2^2$, $\mathfrak{p}_1 \mathfrak{p}_2 \mathfrak{p}_3$ or $\mathfrak{p}_1 \mathfrak{p}_3^2$.

We know that every ideal of $\mathfrak{O}$ is finitely generated. In fact, we shall prove that two generators suffice.

Lemma 5.13. *If* $\mathfrak{a}, \mathfrak{b}$ *are non-zero ideals of* $\mathfrak{O}$ *then there exists* $\alpha \in \mathfrak{a}$ *such that*

$$\alpha\mathfrak{a}^{-1} + \mathfrak{b} = \mathfrak{O}.$$

Proof. First note that if $\alpha \in \mathfrak{a}$ we have $\mathfrak{a} \mid \alpha$ so that $\alpha\mathfrak{a}^{-1}$ is an ideal and not just a fractional ideal. Now $\alpha\mathfrak{a}^{-1} + \mathfrak{b}$ is the greatest common divisor of $\alpha\mathfrak{a}^{-1}$ and $\mathfrak{b}$, so it is sufficient to choose $\alpha \in \mathfrak{a}$ so that

$$\alpha\mathfrak{a}^{-1} + \mathfrak{p}_i = \mathfrak{O} \qquad (i = 1, \ldots, r)$$

where $\mathfrak{p}_1, \ldots, \mathfrak{p}_r$ are the distinct prime ideals dividing $\mathfrak{b}$. This will follow if

$$\mathfrak{p}_i \nmid \alpha\mathfrak{a}^{-1}$$

since $\mathfrak{p}_i$ is a maximal ideal. So it is sufficient to choose $\alpha \in \mathfrak{a}\backslash\mathfrak{a}\mathfrak{p}_i$ for all $i = 1, \ldots, r$.

If $r = 1$ this is easy, for unique factorization of ideals implies $\mathfrak{a} \neq \mathfrak{a}\mathfrak{p}_i$. For $r > 1$ let

$$\mathfrak{a}_i = \mathfrak{a}\mathfrak{p}_1 \ldots \mathfrak{p}_{i-1}\mathfrak{p}_{i+1} \ldots \mathfrak{p}_r.$$

By the case $r = 1$ we can choose

$$\alpha_i \in \mathfrak{a}_i\backslash\mathfrak{a}_i\mathfrak{p}_i.$$

Define

$$\alpha = \alpha_1 + \ldots + \alpha_r.$$

Then each $\alpha_i \in \mathfrak{a}_i \subseteq \mathfrak{a}$, so $\alpha \in \mathfrak{a}$. Suppose if possible that $\alpha \in \mathfrak{a}\mathfrak{p}_i$. If $j \neq i$ then $\alpha_j \in \mathfrak{a}_j \subseteq \mathfrak{a}\mathfrak{p}_i$, so it follows that

$$\alpha_i = \alpha - \alpha_1 - \ldots - \alpha_{i-1} - \alpha_{i+1} - \ldots - \alpha_r \in \mathfrak{a}\mathfrak{p}_i.$$

Hence $\mathfrak{a}\mathfrak{p}_i \mid \langle\alpha_i\rangle$. On the other hand $\mathfrak{a}_i \mid \langle\alpha_i\rangle$. We have $\mathfrak{a}_i\mathfrak{p}_i \mid \langle\alpha_i\rangle$.
This contradicts the choice of α_i. □

Theorem 5.14. *Let* $\mathfrak{a} \neq 0$ *be an ideal of* $\mathfrak{O}$, *and* $0 \neq \beta \in \mathfrak{a}$. *Then there exists* $\alpha \in \mathfrak{a}$ *such that* $\mathfrak{a} = \langle\alpha, \beta\rangle$.

Proof. Let $\mathfrak{b} = \beta\mathfrak{a}^{-1}$. By Lemma 5.13 there exists $\alpha \in \mathfrak{a}$ such that

$$\alpha\mathfrak{a}^{-1} + \mathfrak{b} = \alpha\mathfrak{a}^{-1} + \beta\mathfrak{a}^{-1} = \mathfrak{O},$$

hence

$$(\langle\alpha\rangle + \langle\beta\rangle)\mathfrak{a}^{-1} = \mathfrak{O},$$

so that

$$\mathfrak{a} = \langle\alpha\rangle + \langle\beta\rangle = \langle\alpha, \beta\rangle.$$

□

This theorem demonstrates that the experience of the earlier extended example, where each ideal considered had at most two generators, is entirely typical of ideals in a ring of integers of a number field.

We are now in a position to characterize those $\mathfrak{O}$ for which factorization of elements into irreducibles is unique:

Theorem 5.15. *Factorization of elements of $\mathfrak{O}$ into irreducibles is unique if and only if every ideal of $\mathfrak{O}$ is principal.*

Proof. If every ideal is principal, then unique factorization of elements follows by Theorem 4.15. To prove the converse, if factorization of elements is unique, it will be sufficient to prove that every *prime* ideal is principal, since every other ideal, being a product of prime ideals, would then be principal. Let $\mathfrak{p} \neq 0$ be a prime ideal of $\mathfrak{O}$. By Theorem 5.11(b) there exists a rational integer $N = \mathrm{N}(\mathfrak{p})$ such that $\mathfrak{p} \mid N$. We can factorize N as a product of irreducible elements in $\mathfrak{O}$, say

$$N = \pi_1 \ldots \pi_s.$$

Since $\mathfrak{p} \mid N$ and $\mathfrak{p}$ is a prime ideal, it follows that $\mathfrak{p} \mid \pi_i$, or equivalently, $\mathfrak{p} \mid \langle\pi_i\rangle$. But factorization being unique in $\mathfrak{O}$, the irreducible π_i is actually *prime* by Theorem 4.13, and then the principal ideal $\langle\pi_i\rangle$ is prime (Exercise 5.1). Thus $\mathfrak{p} \mid \langle\pi_i\rangle$, where both $\mathfrak{p}$, $\langle\pi_i\rangle$ are prime, and by uniqueness of factorization,

$$\mathfrak{p} = \langle\pi_i\rangle,$$

so $\mathfrak{p}$ is principal. □

Using this theorem we can nicely round off the relationship

between factorization of elements and ideals. To do this, consider an element π which is irreducible but not prime. Then the ideal $\langle\pi\rangle$ is not prime, so has a proper factorization into prime ideals:

$$\langle\pi\rangle = \mathfrak{p}_1 \ldots \mathfrak{p}_r.$$

Now none of these $\mathfrak{p}_i$ can be principal, for if $\mathfrak{p}_i = \langle a\rangle$, then $\langle a\rangle \mid \langle\pi\rangle$, implying $a \mid \pi$. Since π is irreducible, a would either be a unit (contradicting $\langle a\rangle$ prime), or an associate of π, whence $\langle\pi\rangle = \mathfrak{p}_i$, contradicting the fact that $\langle\pi\rangle$ has a proper factorization.

Tying up the loose ends, we see that if $\mathfrak{O}$ has unique factorization of elements into irreducibles, then these irreducibles are all primes; and factorization of elements corresponds precisely to factorization of the corresponding principal ideals. On the other hand, if $\mathfrak{O}$ does not have unique factorization of elements, then not all irreducibles are prime, and any non-prime irreducible generates a principal ideal which has a proper factorization into non-principal ideals. We may add in the latter case that such non-principal ideals have precisely two generators.

Example. In $\mathbf{Z}[\sqrt{-17}]$, the elements 2, 3 are irreducible (proved by considering norms) and not prime, with

$$\langle 2\rangle = \langle 2, 1+\sqrt{-17}\rangle^2$$
$$\langle 3\rangle = \langle 3, 1+\sqrt{-17}\rangle\langle 3, 1-\sqrt{-17}\rangle.$$

5.4 Nonunique factorization in cyclotomic fields

We mentioned in the introductory section that unique prime factorization *fails* in the cyclotomic field of 23rd roots of unity. (The failure, rather than the precise value $n = 23$, is the crucial point!) In this (optional) section we use the tools developed in this chapter to demonstrate this result. The calculations are somewhat tedious and have been abbreviated where feasible: the energetic reader may care to check the details. A few tricks, inspired by the structure of the group of units of the ring $\mathbf{Z}_{23}$, are used; but we lack the space to motivate them. For further details see the admirable book by Edwards [4].

Let $\zeta = e^{2\pi i/23}$, and let $K = \mathbf{Q}(\zeta)$. By Theorem 3.5 we know that the ring of integers $\mathfrak{O}_K$ is $\mathbf{Z}[\zeta]$. The group of units of $\mathbf{Z}_{23}$ is generated by -2, whose powers in order are

$$1, -2, 4, -8, -7, 14, -5, \ldots . \tag{11}$$

For reasons that will emerge later we introduce two elements

$$\begin{aligned}\theta_0 &= \zeta + \zeta^4 + \zeta^{-7} + \zeta^{-5} + \ldots,\\ \theta_1 &= \zeta^{-2} + \zeta^{-8} + \zeta^{14} + \ldots .\end{aligned}$$

The powers that occur are alternate elements in the sequence (11). We have

$$\begin{aligned}\theta_0 + \theta_1 &= \zeta + \zeta^2 + \ldots + \zeta^{22} = -1,\\ \theta_0\theta_1 &= 6.\end{aligned}$$

The norm of a general element $f(\zeta)$, with f a polynomial over $\mathbf{Z}$ of degree $\leqslant 22$, can be broken up as

$$\begin{aligned}N(f(\zeta)) &= \prod_{j=1}^{22} Nf(\zeta^j)\\ &= \prod_{j \text{ even}} Nf(\zeta^j) \cdot \prod_{j \text{ odd}} Nf(\zeta^j)\\ &= G(\zeta^2)G(\zeta^{-2})\end{aligned} \tag{10}$$

where

$$G(\zeta) = f(\zeta)f(\zeta^4)f(\zeta^{-7})f(\zeta^{-5})f(\zeta^3)f(\zeta^{-11})f(\zeta^2)f(\zeta^8)f(\zeta^9)$$
$$f(\zeta^{-10})f(\zeta^6).$$

By definition, $G(\zeta)$ is invariant under the linear mapping α sending ζ^j to ζ^{4j}. But it is easy to check that an element fixed by α must be of the form $a + b\theta_0$ for $a, b \in \mathbf{Z}$. (Either use Galois theory, or a direct argument based on the linear independence over $\mathbf{Q}$ of $\{1, \zeta, \ldots, \zeta^{22}\}$.)

We pull out of a hat the element

$$\mu = 1 - \zeta + \zeta^{21} = 1 - \zeta + \zeta^{-2},$$

which Kummer found by a great deal of (fairly systematic) experimentation. Using (11) above and a lot of paper and ink we eventually find that

$$N(\mu) = (-31 + 28\theta_0)(-31 + 28\theta_1) = 6533 = 47.139.$$

By Theorem 5.11 the principal ideal $\mathfrak{m} = \langle\mu\rangle$ cannot be prime, hence it must be nontrivial product of prime ideals, say

$$\mathfrak{m} = \mathfrak{p}\mathfrak{q}\,.$$

Taking norms we must (without loss of generality) have $N(\mathfrak{p}) = 47, N(\mathfrak{q}) = 139$. If K has unique factorization then every ideal, in particular $\mathfrak{p}$, is principal by Theorem 5.15. Hence $\mathfrak{p} = \langle\nu\rangle$ for some $\nu \in \mathbf{Z}[\zeta]$. Clearly $N(\nu) = \pm 47$ by Corollary 5.9.

We claim this is impossible. We have already observed that $G(\zeta)$ can always be expressed in the form $a + b\theta_0$ $(a, b, \in \mathbf{Z})$; and then $G(\zeta^{-2})$ must be equal to $a + b\theta_1$. Hence, setting $f(\zeta) = \nu$, we get

$$\pm 47 = (a + b\theta_0)(a + b\theta_1) = a^2 - ab + 6b^2.$$

Multiplying by 4 and regrouping, we find that

$$(2a-b)^2 + 23b^2 = \pm 188.$$

The sign must be positive. A simple trial-and-error analysis (involving only two cases) shows that 188 cannot be written in the form $P^2 + 23Q^2$. This contradiction establishes that prime factorization of elements cannot be unique in $\mathfrak{O}_K$.

Exercises

5.1 In an integral domain D, show that a principal ideal $\langle p \rangle$ is prime if and only if p is a prime or zero.

5.2 In $\mathbf{Z}[\sqrt{-5}]$, define the ideals

$$\begin{aligned} \mathfrak{p} &= \langle 2, 1+\sqrt{-5} \rangle, \\ \mathfrak{q} &= \langle 3, 1+\sqrt{-5} \rangle, \\ \mathfrak{r} &= \langle 3, 1-\sqrt{-5} \rangle. \end{aligned}$$

Prove that these are maximal ideals, hence prime. Show that

$$\begin{aligned} \mathfrak{p}^2 &= \langle 2 \rangle, & \mathfrak{q}\mathfrak{r} &= \langle 3 \rangle, \\ \mathfrak{p}\mathfrak{q} &= \langle 1+\sqrt{-5} \rangle, & \mathfrak{p}\mathfrak{r} &= \langle 1-\sqrt{-5} \rangle. \end{aligned}$$

Show that the factorizations of 6 given in the proof of Theorem 4.10 come from two different groupings of the factorization into prime ideals $\langle 6 \rangle = \mathfrak{p}^2\mathfrak{q}\mathfrak{r}$.

5.3 Calculate the norms of the ideals mentioned in Exercise 5.2 and check multiplicativity.

5.4 Prove that the ideals $\mathfrak{p}$, $\mathfrak{q}$, $\mathfrak{r}$ of Exercise 5.2 cannot be principal.

5.5 Show the principal ideals $\langle 2 \rangle$, $\langle 3 \rangle$ in Exercise 5.2 are generated by irreducible elements but the ideals are not prime.

5.6 In $\mathbf{Z}[\sqrt{-6}]$ we have

$$6 = 2.3 = (\sqrt{-6})(-\sqrt{-6}).$$

Factorize these elements further in the extension ring $\mathbf{Z}[\sqrt{2}, \sqrt{-3}]$ as

$$6 = (-1)\sqrt{2}\sqrt{2}\sqrt{-3}\sqrt{-3}.$$

Show that if $\mathfrak{I}_1$ is the principal ideal in $\mathbf{Z}[\sqrt{2}, \sqrt{-3}]$

generated by $\sqrt{2}$, then

$$\mathfrak{p}_1 = \mathfrak{I}_1 \cap \mathbf{Z}[\sqrt{-6}] = \langle 2, \sqrt{-6}\rangle.$$

Demonstrate that $\mathfrak{p}_1$ is maximal in $\mathbf{Z}[\sqrt{-6}]$, hence prime; and find another prime ideal $\mathfrak{p}_2$ in $\mathbf{Z}[\sqrt{-6}]$ such that

$$\langle 6\rangle = \mathfrak{p}_1^2 \mathfrak{p}_2^2.$$

5.7 Factorize $14 = 2.7 = (2+\sqrt{-10})(2-\sqrt{-10})$ further in $\mathbf{Z}[\sqrt{-5}, \sqrt{2}]$ and by intersecting appropriate ideals with $\mathbf{Z}[\sqrt{-10}]$, factorize the ideal $\langle 14\rangle$ into prime (maximal) ideals in $\mathbf{Z}[\sqrt{-10}]$.

5.8 Suppose $\mathfrak{p}$, $\mathfrak{q}$ are distinct prime ideals in $\mathfrak{O}$. Show $\mathfrak{p}+\mathfrak{q} = \mathfrak{O}$ and $\mathfrak{p} \cap \mathfrak{q} = \mathfrak{p}\mathfrak{q}$.

5.9 If $\mathfrak{O}$ is a principal ideal domain, prove that every fractional ideal is of the form $\{\alpha\phi \mid \alpha \in \mathfrak{O}\}$ for some $\phi \in K$. Does the converse hold?

5.10 Find all fractional ideals of $\mathbf{Z}$ and of $\mathbf{Z}[\sqrt{-1}]$.

5.11 In $\mathbf{Z}[\sqrt{-5}]$, find a $\mathbf{Z}$-basis $\{\alpha_1, \alpha_2\}$ for the ideal $\langle 2, 1+\sqrt{-5}\rangle$. Check the formula

$$N(\langle 2, 1+\sqrt{-5}\rangle) = \left|\frac{\Delta[\alpha_1, \alpha_2]}{d}\right|^{1/2}$$

of Theorem 5.8.

5.12 Find all the ideals in $\mathbf{Z}[\sqrt{-5}]$ which contain the element 6.

5.13 Find all the ideals in $\mathbf{Z}[\sqrt{2}]$ with norm 18.

5.14 If K is a number field of degree n with integers $\mathfrak{O}$, show that if $m \in \mathbf{Z}$ and $\langle m\rangle$ is the ideal in $\mathfrak{O}$ generated by m, then

$$\mathrm{N}(\langle m\rangle) = |m|^n.$$

5.15 In $\mathbf{Z}[\sqrt{-29}]$ we have

$$30 = 2.3.5 = (1+\sqrt{-29})(1-\sqrt{-29})$$

Show

$$\langle 30\rangle \subseteq \langle 2, 1+\sqrt{-29}\rangle$$

and verify $\mathfrak{p}_1 = \langle 2, 1+\sqrt{-29}\rangle$ has norm 2 and is thus prime. Check that $1-\sqrt{-29} \in \mathfrak{p}_1$ and deduce $\langle 30\rangle \subseteq \mathfrak{p}_1^2$. Find prime ideals $\mathfrak{p}_2, \mathfrak{p}_2', \mathfrak{p}_3, \mathfrak{p}_3'$ with norms 3 or 5 such that

$$\langle 30\rangle \subseteq \mathfrak{p}_i\mathfrak{p}_i' \ (i = 2, 3).$$

Deduce that $\mathfrak{p}_1^2\mathfrak{p}_2\mathfrak{p}_2'\mathfrak{p}_3\mathfrak{p}_3' \mid \langle 30\rangle$ and by calculating norms, or otherwise, show that

$$\langle 30\rangle = \mathfrak{p}_1^2\mathfrak{p}_2\mathfrak{p}_2'\mathfrak{p}_3\mathfrak{p}_3'.$$

Comment on how this relates to the two factorizations:

$$\langle 30\rangle = \langle 2\rangle\langle 3\rangle\langle 5\rangle,$$
$$\langle 30\rangle = \langle 1+\sqrt{-29}\rangle\langle 1-\sqrt{-29}\rangle.$$

5.16 Find all ideals in $\mathbf{Z}[\sqrt{-29}]$ containing the element 30.

PART II

Geometric methods

CHAPTER SIX

Lattices

At this stage in the theory purely algebraic methods start to be supplanted by geometric techniques, requiring a different attitude of mind on the part of the reader. (At a more advanced stage of the theory, which we never reach, methods from complex analysis make an appearance; and from this point on all three approaches are used simultaneously.) Here we develop some properties of lattices: subsets of $\mathbf{R}^n$ which in some sense generalize the way $\mathbf{Z}$ is embedded in $\mathbf{R}$. We characterize lattices topologically as the discrete subgroups of $\mathbf{R}^n$. We introduce the fundamental domain and quotient torus corresponding to a lattice and relate the two concepts. Finally we define a concept of volume for subsets of the quotient torus.

6.1 Lattices

Let $e_1, \ldots, e_m$ be a linearly independent set of vectors in $\mathbf{R}^n$ (so that $m \leqslant n$). The additive subgroup of $(\mathbf{R}^n, +)$ generated by $e_1, \ldots, e_m$ is called a *lattice* of *dimension m, generated* by $e_1, \ldots, e_m$. Fig. 1 shows a lattice of dimension 2 in $\mathbf{R}^2$, generated by $(1, 2)$ and $(2, -1)$. (Do not confuse this with any other uses of the word 'lattice' in algebra.) Obviously, as regards the group-theoretic structure, a lattice

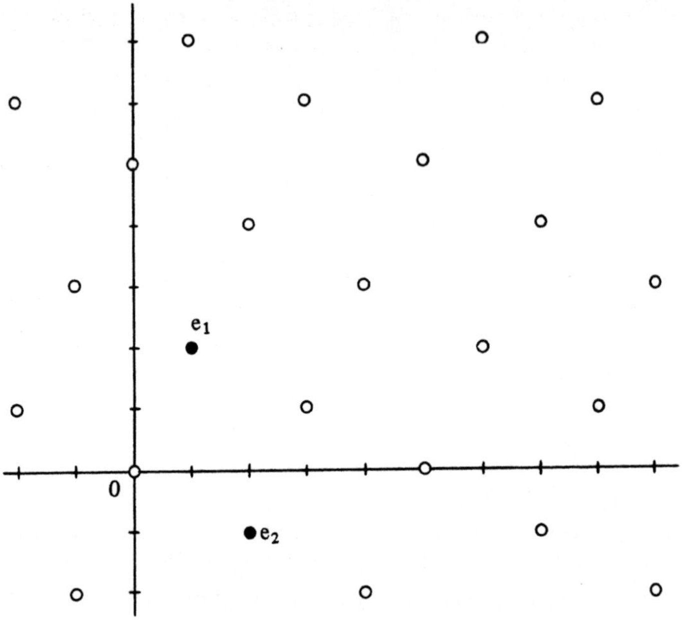

Fig. 1. The lattice in $\mathbf{R}^2$ generated by $e_1 = (1, 2)$ and $e_2 = (2, -1)$.

of dimension m is a free abelian group of rank m, so we can apply the terminology and theory of free abelian groups to lattices.

We shall give a topological characterization of lattices. Let $\mathbf{R}^n$ be equipped with the usual metric (à la Pythagoras), where $\|x - y\|$ denotes the distance between x and y, and denote the (closed) ball centre x radius r by $B_r[x]$. Recall that a subset $X \subseteq \mathbf{R}^n$ is *bounded* if $X \subseteq B_r[0]$ for some r. We say that a subset of $\mathbf{R}^n$ is *discrete* if and only if it intersects every $B_r[0]$ in a finite set.

Theorem 6.1. *An additive subgroup of* $\mathbf{R}^n$ *is a lattice if and only if it is discrete.*

Proof. Suppose L is a lattice. By passing to the subspace spanned by L we may assume L has dimension n. Let L be generated by $e_1, \ldots, e_n$; then these vectors form a basis for the space $\mathbf{R}^n$. Every $v \in \mathbf{R}^n$ has a unique representation

$$v = \lambda_1 e_1 + \ldots + \lambda_n e_n \qquad (\lambda_i \in \mathbf{R}).$$

Define $f : \mathbf{R}^n \to \mathbf{R}^n$ by

$$f(\lambda_1 e_1 + \ldots + \lambda_n e_n) = (\lambda_1, \ldots, \lambda_n).$$

Then $f(B_r[0])$ is bounded, say

$$\|f(v)\| \leqslant k \text{ for } v \in B_r[0].$$

If $\Sigma a_i e_i \in B_r[0]$ $(a_i \in \mathbf{Z})$, then certainly $\|(a_1, \ldots, a_n)\| \leqslant k$. This implies

$$|a_i| \leqslant \|(a_1, \ldots, a_n)\| \leqslant k. \tag{1}$$

The number of integer solutions of (1) is finite and so $L \cap B_r[0]$, being a subset of the solutions of (1), is also finite, and L is discrete.

Conversely, let G be a discrete subgroup of $\mathbf{R}^n$. We prove by induction on n that G is a lattice. Let $\{g_1, \ldots, g_m\}$ be a maximal linearly independent subset of G, let V be the subspace spanned by $\{g_1, \ldots, g_{m-1}\}$, and let $G_0 = G \cap V$. Then G_0 is discrete so by induction is a lattice. Hence there exist linearly independent elements $h_1, \ldots, h_{m'}$ generating G_0. Since the elements $g_1, \ldots, g_{m-1} \in G_0$ we have $m' = m - 1$, and we can replace $\{g_1, \ldots, g_{m-1}\}$ by $\{h_1, \ldots, h_{m-1}\}$, or equivalently assume that every element of G_0 is a $\mathbf{Z}$-linear combination of $g_1, \ldots, g_{m-1}$. Let T be the subset of all $x \in G$ of the form

$$x = a_1 g_1 + \ldots + a_m g_m$$

with $a_i \in \mathbf{R}$, such that

$$0 \leqslant a_i < 1 \qquad (i = 1, \ldots, m-1)$$
$$0 \leqslant a_m \leqslant 1.$$

Then T is bounded, hence finite since G is discrete, and we may therefore choose $x' \in T$ with smallest non-zero coefficient a_m, say

$$x' = b_1 g_1 + \ldots + b_m g_m.$$

Certainly $\{g_1, \ldots, g_{m-1}, x'\}$ is linearly independent. Now starting with any vector $g \in G$ we can select integer coefficients c_i so that

$$g' = g - c_m x' - c_1 g_1 - \ldots - c_{m-1} g_{m-1}$$

lies in T, and the coefficient of g_m in g' is less than b_m, but non-negative. By choice of x' this coefficient must be zero, so $g' \in G_0$. Hence $\{x', g_1, \ldots, g_{m-1}\}$ generates G, and G is a lattice. □

If L is a lattice generated by $\{e_1, \ldots, e_n\}$ we define the *fundamental domain* T to consist of all elements $\Sigma\, a_i e_i$ ($a_i \in \mathbf{R}$) for which

$$0 \leqslant a_i < 1.$$

Note that this depends on the choice of generators.

Lemma 6.2. *Each element of* $\mathbf{R}^n$ *lies in exactly one of the sets* $T + l$ *for* $l \in L$.

Proof. Chop off the integer parts of the coefficients. □

Fig. 2 illustrates the concept of a fundamental domain, and the result of Lemma 6.2, for the lattice of Fig. 1.

6.2 The quotient torus

Let L be a lattice in $\mathbf{R}^n$, and assume to start with that L has dimension n. We shall study the quotient group $\mathbf{R}^n/L$.

Let $\mathbf{S}$ denote the set of all complex numbers of modulus 1. Under multiplication $\mathbf{S}$ is a group, called for obvious reasons the *circle group*.

Lemma 6.3. *The quotient group* $\mathbf{R}/\mathbf{Z}$ *is isomorphic to the circle group* $\mathbf{S}$.

Proof. Define a map $\phi : \mathbf{R} \to \mathbf{S}$ by

$$\phi(x) = e^{2\pi i x}.$$

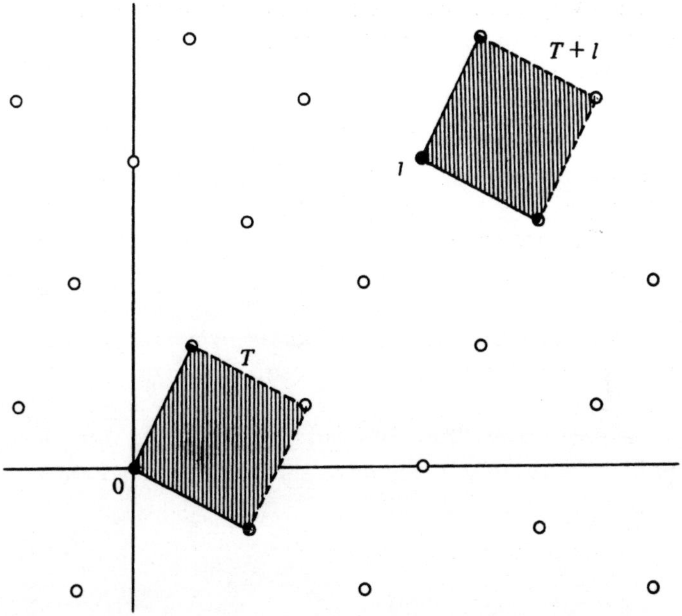

Fig. 2. A fundamental domain T for the lattice of Fig. 1, and a translate $T + l$. Dotted lines indicate omission of boundaries.

Then ϕ is a surjective homomorphism with kernel $\mathbf{Z}$, and the lemma follows. □

Next let $\mathbf{T}^n$ denote the direct product of n copies of $\mathbf{S}$, and call this the *n-dimensional torus*. For instance, $\mathbf{T}^2 = \mathbf{S} \times \mathbf{S}$ is the usual torus (with a group structure) as sketched in Fig. 3.

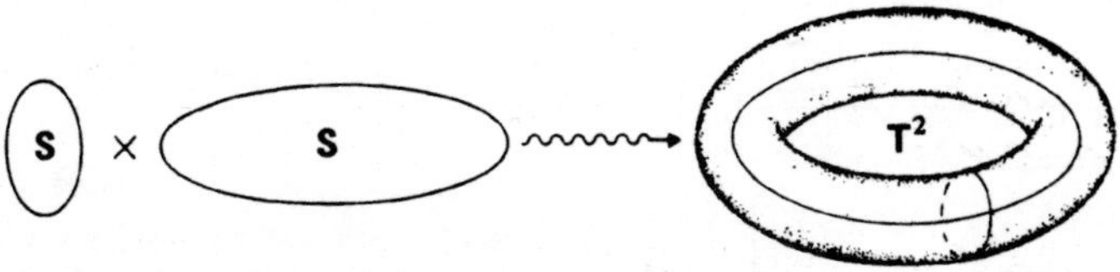

Fig. 3. The Cartesian product of two circles is a torus.

Theorem 6.4. *If L is an n-dimensional lattice in* $\mathbf{R}^n$ *then* $\mathbf{R}^n/L$ *is isomorphic to the n-dimensional torus* $\mathbf{T}^n$.

Proof. Let $\{e_1, \ldots, e_n\}$ be generators for L. Then $\{e_1, \ldots, e_n\}$ is a basis for $\mathbf{R}^n$. Define $\phi: \mathbf{R}^n \to \mathbf{T}^n$ by

$$\phi(a_1e_1 + \ldots + a_ne_n) = (e^{2\pi i a_1}, \ldots, e^{2\pi i a_n}).$$

Then ϕ is a surjective homorphism, and the kernel of ϕ is L. □

Lemma 6.5. *The map ϕ defined above, when restricted to the fundamental domain T, yields a bijection $T \to \mathbf{T}^n$.* □

Geometrically, $\mathbf{T}^n$ is obtained by 'glueing' (i.e. identifying) opposite faces of the closure of the fundamental domain, as in Fig. 4.

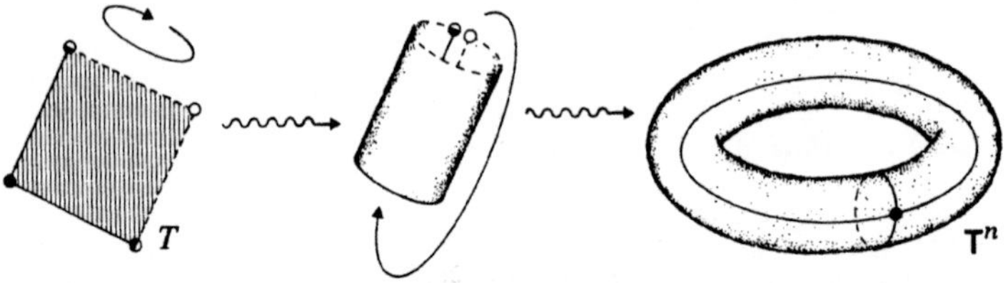

Fig. 4. The quotient of Euclidean space by a lattice of the same dimension is a torus, obtained by identifying opposite edges of a fundamental domain.

If the dimension of L is less than n, we have a similar result:

Theorem 6.6. *Let L be an m-dimensional lattice in* $\mathbf{R}^n$. *Then* $\mathbf{R}^n/L$ *is isomorphic to* $\mathbf{T}^m \times \mathbf{R}^{n-m}$.

Proof. Let V be the subspace spanned by L, and choose a complement W so that $\mathbf{R}^n = V \oplus W$. Then $L \subseteq V$, $V/L \cong \mathbf{T}^m$ by Theorem 6.4, $W \cong \mathbf{R}^{n-m}$, and the result follows. □

For example, $\mathbf{R}^2/\mathbf{Z} \cong \mathbf{T}^1 \times \mathbf{R}$, which geometrically is a cylinder as in Fig. 5.

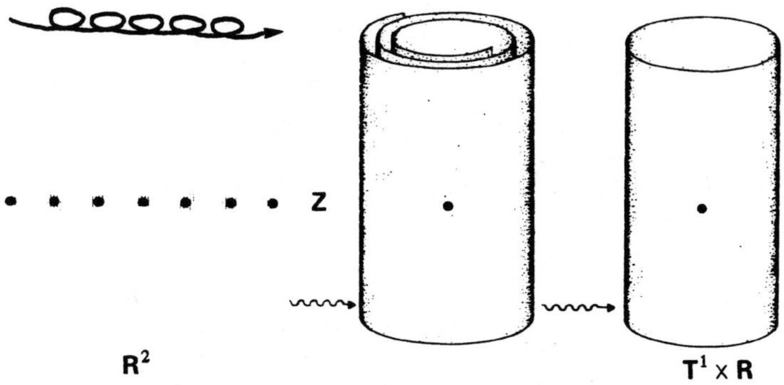

Fig. 5. The quotient of Euclidean space by a lattice of smaller dimension is a cylinder.

The *volume* $v(X)$ of a subset $X \subseteq \mathbf{R}^n$ is defined in the usual way: for precision we take it to be the value of the multiple integral

$$\int_X \mathrm{d}x_1 \ldots \mathrm{d}x_n$$

where $(x_1, \ldots, x_n)$ are co-ordinates. Of course the volume exists only when the integral does.

Let $L \subseteq \mathbf{R}^n$ be a lattice of dimension n, so that $\mathbf{R}^n/L \cong \mathbf{T}^n$. Let T be a fundamental domain of L. We have noted the existence of a bijection

$$\phi : T \to \mathbf{T}^n.$$

For any subset X of $\mathbf{T}^n$ we define the *volume* $v(X)$ by

$$v(X) = v(\phi^{-1}(X))$$

which exists if and only if $\phi^{-1}(X)$ has a volume in $\mathbf{R}^n$.

Let $\nu : \mathbf{R}^n \to \mathbf{T}^n$ be the natural homomorphism with kernel L. It is intuitively clear that ν is 'locally volume-preserving', that is, for each $x \in \mathbf{R}^n$ there exists a ball $B_\epsilon[x]$ such that for all subsets $X \subseteq B_\epsilon[x]$ for which $v(X)$ exists we have

$$v(X) = v(\nu(X))$$

It is also intuitively clear that if an injective map is locally volume-preserving then it is volume-preserving. We prove a result which combines these two intuitive ideas:

Theorem 6.7. *If X is a bounded subset of* $\mathbf{R}^n$ *and $v(X)$ exists, and if $v(\nu(X)) \neq v(X)$, then $\nu|_X$ is not injective.*

Proof. Assume $\nu|_X$ is injective. Now X, being bounded, intersects only a finite number of the sets $T + l$, for T a fundamental domain and $l \in L$. Put

$$X_l = X \cap (T + l).$$

Then we have

$$X = X_{l_1} \cup \ldots \cup X_{l_n}$$

say. For each l_i define

$$Y_{l_i} = X_{l_i} - l_i,$$

so that $Y_{l_i} \subseteq T$. We claim that the Y_{l_i} are disjoint. Since $\nu(x - l_i) = \nu(x)$ for all $x \in \mathbf{R}^n$ this follows from the assumed injectivity of ν. Now

$$v(X_{l_i}) = v(Y_{l_i})$$

for all i. Also

$$\nu(X_{l_i}) = \phi(Y_{l_i})$$

where ϕ is the bijection $T \to \mathbf{T}^n$. Now we compute:

$$\begin{aligned} v(\nu(X)) &= v(\nu(\cup X_{l_i})) \\ &= v(\cup Y_{l_i}) \\ &= \sum v(Y_{l_i}) \quad \text{by disjointness} \\ &= \sum v(X_{l_i}) \\ &= v(X), \end{aligned}$$

which is a contradiction. □

The idea of the proof can be summed up pictorially by Fig. 6.

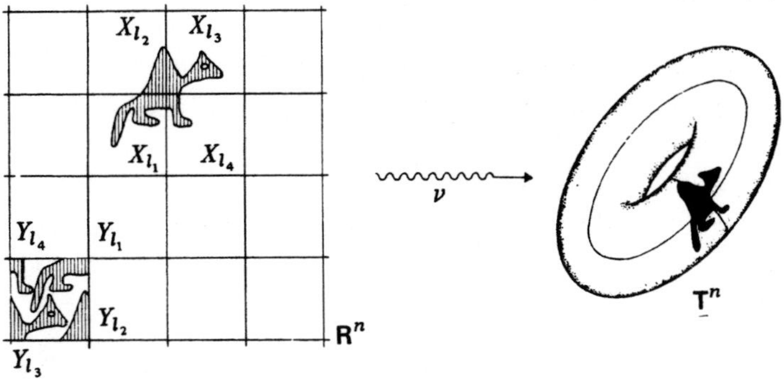

Fig. 6. Proof of Theorem 6.7: if a locally volume-preserving map does not preserve volume globally, then it cannot be injective.

Exercises

6.1 Let L be a lattice in $\mathbf{R}^2$ with $L \subseteq \mathbf{Z}^2$. Prove that the volume of a fundamental domain T is equal to the number of points of $\mathbf{Z}^2$ lying in T.

6.2 Generalize the previous exercise to $\mathbf{R}^n$ and link this to Lemma 9.3 by using Theorem 1.13.

6.3 Sketch the lattices in $\mathbf{R}^2$ generated by:

(a) (0, 1) and (1, 0).
(b) (− 1, 2) and (2, 2).
(c) (1, 1) and (2, 3).
(d) (− 2, − 7) and (4, − 3).
(e) (1, 20) and (1, − 20).
(f) $(1, \pi)$ and $(\pi, 1)$.

6.4 Sketch fundamental domains for these lattices.

6.5 Hence show that the fundamental domain of a lattice is not uniquely determined until we specify a set of generators.

6.6 Verify that nonetheless the volume of a fundamental domain of a given lattice is independent of the set of generators chosen.

6.7 Find two different fundamental domains for the lattice in $\mathbf{R}^3$ generated by (0, 0, 1), (0, 2, 0), (1, 1, 1). Show by direct calculation that they have the same volume. Can you prove this geometrically by dissecting the fundamental domains into mutually congruent pieces?

CHAPTER SEVEN

Minkowski's theorem

The aim of this chapter is to prove a marvellous theorem, due to Minkowski in 1896. This asserts the existence within a suitable set X of a non-zero point of a lattice L, provided the volume of X is sufficiently large relative to that of a fundamental domain of L. The idea behind the proof is deceptive in its simplicity: it is that X cannot be squashed into a space whose volume is less than that of X, unless X is allowed to overlap itself. Minkowski discovered that this essentially trivial observation has many non-trivial and important consequences, and used it as a foundation for an extensive theory of the 'geometry of numbers'. As immediate and accessible instances of its application we prove the two- and four-squares theorems of classical number theory.

7.1 Minkowski's theorem

A subset $X \subseteq \mathbf{R}^n$ is *convex* if whenever $x, y \in X$ then all points on the straight line segment joining x to y also lie in X. In algebraic terms, X is convex if, whenever $x, y \in X$, the point

$$\lambda x + (1 - \lambda) y$$

belongs to x for all real λ, $0 \leqslant \lambda \leqslant 1$.

For example a circle, a square, an ellipse, or a triangle is convex in $\mathbf{R}^2$, but an annulus or crescent is not (Fig. 7). A subset $X \subseteq \mathbf{R}^n$ is (*centrally*) *symmetric* if $x \in X$ implies $-x \in X$. Geometrically this means that X is invariant under reflection in the origin. Of the sets in Fig. 7, assuming the origin to be at the positions marked with an asterisk, the circle, square, ellipse, and annulus are symmetric, but the triangle and crescent are not.

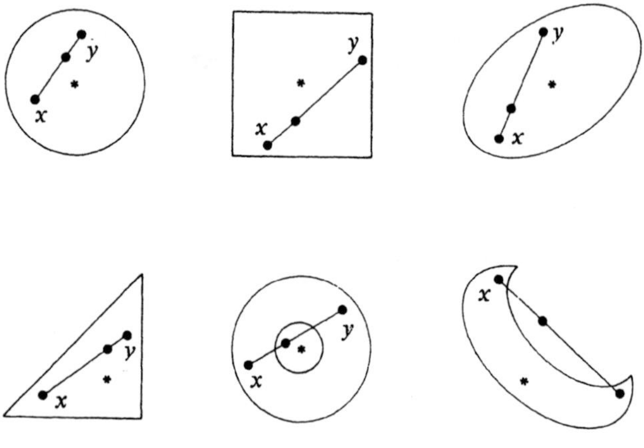

Fig. 7. Convex and non-convex sets. The circular disc, square, ellipse, and triangle are convex; the annulus and crescent are not. The circle, square, ellipse, and annulus are centrally symmetric about *; the triangle and crescent are not.

We may now state Minkowski's theorem.

Theorem 7.1. (*Minkowski's theorem*) *Let L be an n-dimensional lattice in $\mathbf{R}^n$ with fundamental domain T, and let X be a bounded symmetric convex subset of $\mathbf{R}^n$. If*

$$v(X) > 2^n v(T)$$

then X contains a non-zero point of L.

Proof. Double the size of L to obtain a lattice $2L$ with fundamental domain $2T$ of volume $2^n v(T)$. Consider the torus

$$\mathbf{T}^n = \mathbf{R}^n/2L.$$

By definition,

$$v(\mathbf{T}^n) = v(2T) = 2^n v(T).$$

Now the natural map $\nu : \mathbf{R}^n \to \mathbf{T}^n$ cannot preserve the volume of X, since this is strictly larger than $v(\mathbf{T}^n)$: since $\nu(X) \subseteq \mathbf{T}^n$ we have

$$v(\nu(X)) \leqslant v(\mathbf{T}^n) = 2^n v(T) < v(X).$$

It follows by Theorem 6.7 that $\nu|_X$ is not injective. Hence there exist $x_1 \neq x_2$, $x_1, x_2 \in X$, such that

$$\nu(x_1) = \nu(x_2),$$

or equivalently

$$x_1 - x_2 \in 2L. \qquad (1)$$

But $x_2 \in X$, so $-x_2 \in X$ by symmetry; and now by convexity

$$\tfrac{1}{2}(x_1) + \tfrac{1}{2}(-x_2) \in X,$$

that is,

$$\tfrac{1}{2}(x_1 - x_2) \in X.$$

But by (1),

$$\tfrac{1}{2}(x_1 - x_2) \in L.$$

Hence

$$0 \neq \tfrac{1}{2}(x_1 - x_2) \in X \cap L,$$

as required. □

The geometrical reasoning is illustrated in Fig. 8. The decisive step in the proof is that since $\mathbf{T}^n$ has smaller volume than X it is impossible to squash X into $\mathbf{T}^n$ without overlap: the ancient platitude of quarts and pint pots. That such olde-worlde wisdom become, in the hands of Minkowski, a weapon of devastating power, was the wonder of the 19th century and a lesson for the 20th. We will unleash this power at several crucial stages in the forthcoming battle. (Note that our original Thespian metaphor has been abandoned in favour of a military one, reinforcing the change of viewpoint from that of the algebraic *voyeur* to that of the geometric participant.) As a more immediate affirmation, we now give two traditional applications to number theory: the 'two-squares' and 'four-squares' theorems.

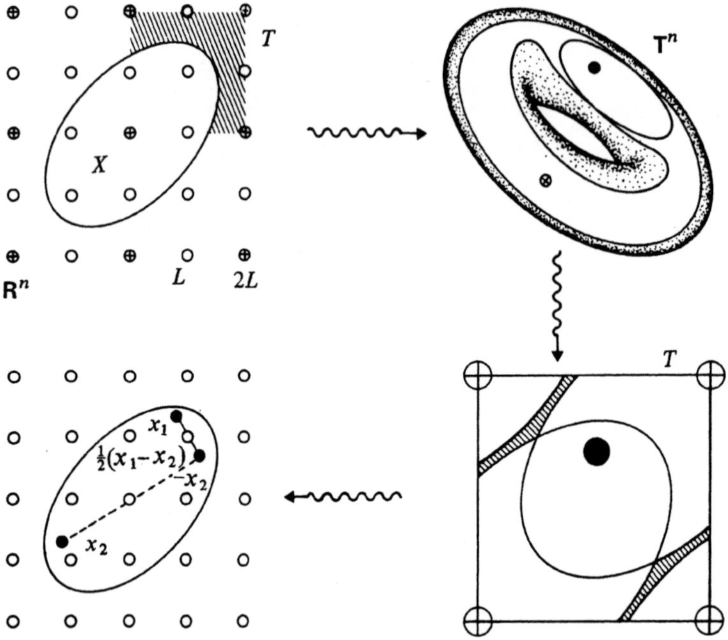

Fig. 8. Proof of Minkowski's theorem. Expand the original lattice (○) to double the size (⊕) and form the quotient torus. By computing volumes, the natural quotient map is not injective when restricted to the given convex set. From points x_1 and x_2 with the same image we may construct a non-zero lattice point $\frac{1}{2}(x_1 - x_2)$.

7.2 The two-squares theorem

We start by proving:

Theorem 7.2. *If p is prime of the form* $4k + 1$ *then p is a sum of two integer squares.*

Proof. The multiplicative group G of the field $\mathbf{Z}_p$ is cyclic ([GT] p. 171) and has order $p - 1 = 4k$. It therefore contains an element u of order 4. Then $u^2 \equiv -1 \pmod{p}$ since -1 is the only element of order 2 in G.

Let $L \subseteq \mathbf{Z}^2$ be the lattice in $\mathbf{R}^2$ consisting of all pairs (a, b) $(a, b \in \mathbf{Z})$ such that

$$b \equiv ua \qquad (\text{mod } p).$$

This is a subgroup of $\mathbf{Z}^2$ of index p (an easy verification left to the reader) so the volume of a fundamental domain for L is p. By Minkowski's theorem any circle, centre the origin, of radius r, which has area

$$\pi r^2 > 4p$$

contains a non-zero point of L. This is the case for $r^2 = 3p/2$. So there exists a point $(a, b) \in L$, not the origin, for which

$$0 \neq a^2 + b^2 \leqslant r^2 = 3p/2 < 2p.$$

But modulo p we have

$$a^2 + b^2 \equiv a^2 + u^2a^2 \equiv 0.$$

Hence $a^2 + b^2$, being a multiple of p strictly between 0 and $2p$, must equal p. □

The reader should draw the lattice L and the relevant circle in a few cases ($p = 5, 13, 17$) and check that the relevant lattice point exists and provides suitable a, b.

Theorem 7.2 goes back to Fermat, who stated it in a letter to Mersenne in 1640. He sent a sketch proof to Pierre de Carcavi in 1659. Euler gave a complete proof in 1754.

7.3 The four-squares theorem

Refining this argument leads to another famous theorem:

Theorem 7.3. *Every positive integer is a sum of four integer squares.*

Proof. We prove the theorem for primes p, and then extend the result to all integers. Now

$$2 = 1^2 + 1^2 + 0^2 + 0^2$$

so we may suppose p is odd. We claim that the congruence

$$u^2 + v^2 + 1 \equiv 0 \qquad (\text{mod } p)$$

has a solution $u, v \in \mathbf{Z}$. This is because u^2 takes exactly $(p+1)/2$ distinct values as u runs through $0, \ldots, p-1$; and $-1-v^2$ also takes on $(p+1)/2$ values: for the congruence to have no solution all these values, $p+1$ in total, will be distinct: then we have $p+1 \leqslant p$ which is absurd.

For such a choice of u, v consider the lattice $L \subseteq \mathbf{Z}^4$ consisting of (a, b, c, d) such that

$$c \equiv ua + vb, \quad d \equiv ub - va \qquad (\mathrm{mod}\ p).$$

Then L has index p^2 in $\mathbf{Z}^4$ (another easy computation) so the volume of a fundamental domain is p^2. Now a 4-dimensional sphere, centre the origin, radius r, has volume

$$\pi^2 r^4/2$$

and we choose r to make this greater than $16p^2$; say $r^2 = 1.9p$.

Then there exists a lattice point $0 \neq (a, b, c, d)$ in this 4-sphere, and so

$$0 \neq a^2 + b^2 + c^2 + d^2 \leqslant r^2 = 1.9p < 2p.$$

Modulo p, it is easy to verify that $a^2 + b^2 + c^2 + d^2 \equiv 0$, hence as before must equal p.

To deal with an arbitrary integer n, it suffices to factorize n into primes and then use the identity

$$\begin{aligned}(a^2+b^2+c^2+d^2)(A^2+B^2+C^2+D^2) \\ = (aA-bB-cC-dD)^2 + (aB+bA+cD-dC)^2 \\ + (aC-bD+cA+dB)^2 + (aD+bC-cB+dA)^2.\end{aligned}$$ □

Theorem 7.3 also goes back to Fermat. Euler spent 40 years trying to prove it, and Lagrange succeeded in 1770.

Exercises

7.1 Which of the following solids are convex? Sphere, pyramid, icosahedron, cube, torus, ellipsoid, parallelepiped?

7.2 How many different convex solids can be made by joining n unit cubes face to face, so that their vertices coincide, for $n = 1, 2, 3, 4, 5, 6$; counting two solids as different if and only if they cannot be mapped to each other by rigid motions? What is the result for general n?

7.3 Verify the two-squares theorem on all primes less than 200.

7.4 Verify the four-squares theorem on all integers less than 100.

7.5 Prove that not every integer is a sum of three squares.

7.6 Prove that the number $\mu(n)$ of pairs of integers (x, y) with $x^2 + y^2 < n$ satisfies $\mu(n)/n \to \pi$ as $n \to \infty$.

CHAPTER EIGHT

Geometric representation of algebraic numbers

The purpose of this chapter is to develop a method of embedding a number field K in a real vector space of dimension equal to the degree of K, in such a way that ideals in K map to lattices in this vector space. This opens the way to applications of Minkowski's theorem. The embedding is defined in terms of the monomorphisms $K \to \mathbf{C}$, and we have to distinguish between those which map K into $\mathbf{R}$ and those which do not.

8.1 The space $\mathbf{L}^{st}$

Let $K = \mathbf{Q}(\theta)$ be a number field of degree n, where θ is an algebraic integer. Let $\sigma_1, \ldots, \sigma_n$ be the set of all monomorphisms $K \to \mathbf{C}$ (see Theorem 2.3). If $\sigma_i(K) \subseteq \mathbf{R}$, which happens if and only if $\sigma_i(\theta) \in \mathbf{R}$, we say that σ_i is *real*; otherwise σ_i is *complex*. As usual denote complex conjugation by bars, and define

$$\bar{\sigma}_i(\alpha) = \overline{\sigma_i(\alpha)}.$$

Since complex conjugation is an automorphism of $\mathbf{C}$ it follows that $\bar{\sigma}_i$ is a monomorphism $K \to \mathbf{C}$, so equals σ_j for some j. Now $\sigma_i = \bar{\sigma}_i$ if and only if σ_i is real, and $\bar{\bar{\sigma}}_i = \sigma_i$, so the complex monomorphisms come in conjugate pairs. Hence

$$n = s + 2t$$

where s is the number of real monomorphisms and $2t$ the number of complex. We standardize the numeration in such a way that the system of all monomorphisms $K \to \mathbf{C}$ is

$$\sigma_1, \ldots, \sigma_s; \sigma_{s+1}, \bar{\sigma}_{s+1}, \ldots, \sigma_{s+t}, \bar{\sigma}_{s+t},$$

where $\sigma_1, \ldots, \sigma_s$ are real and the rest complex.

Further define

$$\mathbf{L}^{st} = \mathbf{R}^s \times \mathbf{C}^t,$$

the set of all $(s + t)$-tuples

$$x = (x_1, \ldots, x_s; x_{s+1}, \ldots, x_{s+t})$$

where $x_1, \ldots, x_s \in \mathbf{R}$ and $x_{s+1}, \ldots, x_{s+t} \in \mathbf{C}$. Then $\mathbf{L}^{st}$ is a vector space over $\mathbf{R}$, and a ring (with co-ordinatewise operations): in fact it is an $\mathbf{R}$-algebra. As vector space over $\mathbf{R}$ it has dimension $s + 2t = n$.

For $x \in \mathbf{L}^{st}$ we define the *norm*

$$\mathrm{N}(x) = x_1 \ldots x_s |x_{s+1}|^2 \ldots |x_{s+t}|^2. \tag{1}$$

(There is no confusion with other uses of the word 'norm', and we shall see why it is desirable to use this apparently overworked word in a moment.) The norm has two obvious properties:

(a) $\mathrm{N}(x)$ is real for all x,

(b) $\mathrm{N}(xy) = \mathrm{N}(x)\mathrm{N}(y)$.

We define a map

$$\sigma : K \to \mathbf{L}^{st}$$

by

$$\sigma(\alpha) = (\sigma_1(\alpha), \ldots, \sigma_s(\alpha); \sigma_{s+1}(\alpha), \ldots, \sigma_{s+t}(\alpha))$$

for $\alpha \in K$. Clearly

$$\sigma(\alpha + \beta) = \sigma(\alpha) + \sigma(\beta) \tag{2}$$

$$\sigma(\alpha\beta) = \sigma(\alpha)\sigma(\beta)$$

for all $\alpha, \beta \in K$; so σ is a ring homomorphism. If r is a rational number then

$$\sigma(r\alpha) = r\sigma(\alpha)$$

so σ is a **Q**-algebra homomorphism. Furthermore, we have

$$N(\sigma(\alpha)) = N(\alpha) \tag{3}$$

since the latter is defined to be

$$\sigma_1(\alpha)\ldots\sigma_s(\alpha)\sigma_{s+1}(\alpha)\bar{\sigma}_{s+1}(\alpha)\ldots\sigma_{s+t}(\alpha)\bar{\sigma}_{s+t}(\alpha)$$

which equals the former.

For example, let $K = \mathbf{Q}(\theta)$ where $\theta \in \mathbf{R}$ satisfies

$$\theta^3 - 2 = 0.$$

Then the conjugates of θ are θ, $\omega\theta$, $\omega^2\theta$ where ω is a complex cube root of unity. The monomorphisms $K \to \mathbf{C}$ are given by

$$\sigma_1(\theta) = \theta, \qquad \sigma_2(\theta) = \omega\theta, \qquad \bar{\sigma}_2(\theta) = \omega^2\theta.$$

Hence $s = t = 1$.

An element of K, say

$$x = q + r\theta + s\theta^2$$

where $q, r, s \in \mathbf{Q}$, maps into $\mathbf{L}^{1,1}$ according to

$$\sigma(x) = (q + r\theta + s\theta^2, q + r\omega\theta + s\omega^2\theta^2).$$

The kernel of σ is an ideal of K since σ is a ring homomorphism. Since K is a field this means that either σ is identically zero or σ is injective. But

$$\sigma(1) = (1, 1, \ldots, 1) \neq 0$$

so σ must be injective. Much stronger is the following result:

Theorem 8.1. *If $\alpha_1, \ldots, \alpha_n$ is a basis for K over* **Q** *then $\sigma(\alpha_1), \ldots, \sigma(\alpha_n)$ are linearly independent over* **R**.

Proof. Linear independence over **Q** is immediate since σ is injective, but we need more than this. Let

$$\sigma_k(\alpha_l) = x_k^{(l)} \qquad (k = 1, \ldots, s)$$

$$\sigma_{s+j}(\alpha_l) = y_j^{(l)} + iz_j^{(l)} \qquad (j = 1, \ldots, t)$$

where $x_k^{(l)}, y_k^{(l)}, z_k^{(l)}$ are real. Then

$$\sigma(\alpha_l) = (x_1^{(l)}, \ldots, x_s^{(l)}; \quad y_1^{(l)} + iz_1^{(l)}, \ldots, y_t^{(l)} + iz_t^{(l)}),$$

and it is sufficient to prove that the determinant

$$D = \begin{vmatrix} x_1^{(1)} \ldots x_s^{(1)} y_1^{(1)} z_1^{(1)} \ldots y_t^{(1)} z_t^{(1)} \\ \ldots \quad \ldots \quad \ldots \\ x_1^{(n)} \ldots x_s^{(n)} y_1^{(n)} z_1^{(n)} \ldots y_t^{(n)} z_t^{(n)} \end{vmatrix}$$

is non-zero. Put

$$E = \begin{vmatrix} x_1^{(1)} \ldots x_s^{(1)} & y_1^{(1)} + iz_1^{(1)} & y_1^{(1)} - iz_1^{(1)} \ldots \\ \ldots & \ldots & \ldots \\ x_1^{(n)} \ldots x_s^{(n)} & y_1^{(n)} + iz_1^{(n)} & y_1^{(n)} - iz_1^{(n)} \ldots \end{vmatrix}$$

$$= \begin{vmatrix} \sigma_1(\alpha_1) \ldots \sigma_s(\alpha_1) & \sigma_{s+1}(\alpha_1) & \bar{\sigma}_{s+1}(\alpha_1) \ldots \\ \ldots & \ldots & \ldots \\ \sigma_1(\alpha_n) \ldots \sigma_s(\alpha_n) & \sigma_{s+1}(\alpha_n) & \bar{\sigma}_{s+1}(\alpha_n) \ldots \end{vmatrix}.$$

Then

$$E^2 = \Delta[\alpha_1, \ldots, \alpha_n]$$

by definition of the discriminant; and by Theorem 2.6, $E^2 \neq 0$. Now elementary properties of determinants (column operations) yield

$$E = (-2i)^t D$$

so that $D \neq 0$ as required. □

Corollary 8.2. *$\mathbf{Q}$-linearly independent elements of K map under σ to $\mathbf{R}$-linearly independent elements of $\mathbf{L}^{st}$.* □

Corollary 8.3. *If G is a finitely generated subgroup of $(K, +)$ with $\mathbf{Z}$-basis $\{\alpha_1, \ldots, \alpha_m\}$ then the image of G in $\mathbf{L}^{st}$ is a lattice with generators $\sigma(\alpha_1), \ldots, \sigma(\alpha_m)$.* □

The 'geometric representation' of K in $\mathbf{L}^{st}$ defined by σ, in combination with Minkowski's theorem, will provide the key to several of the deeper parts of our theory, in Chapters 9, 10, and 12. For these applications we shall need a notion of 'distance' on $\mathbf{L}^{st}$. Since $\mathbf{L}^{st}$, as a real vector space, is isomorphic to $\mathbf{R}^{s+2t}$ the natural thing is to transfer the usual Euclidean metric from $\mathbf{R}^{s+2t}$ to $\mathbf{L}^{st}$. This amounts to choosing a basis in $\mathbf{L}^{st}$ and defining an inner product with respect to which this basis is orthonormal. The natural basis to pick is

$$\left\{\begin{array}{l}(1, 0, \ldots, 0; 0, \ldots, 0) \\ (0, 1, \ldots, 0; 0, \ldots, 0) \\ \qquad\qquad \ldots \\ (0, 0, \ldots, 1; 0, \ldots, 0) \\ (0, 0, \ldots, 0; 1, 0, \ldots, 0) \\ (0, 0, \ldots, 0; \mathrm{i}, 0, \ldots, 0) \\ \qquad\qquad \ldots \\ (0, 0, \ldots, 0; 0, 0, \ldots, 1) \\ (0, 0, \ldots, 0; 0, 0, \ldots, \mathrm{i}).\end{array}\right. \tag{4}$$

With respect to this basis the element

$$(x_1, \ldots, x_s; y_1 + iz_1, \ldots, y_t + iz_t)$$

of $\mathbf{L}^{st}$ has co-ordinates

$$(x_1, \ldots, x_s, y_1, z_1, \ldots, y_t, z_t).$$

Changing notation slightly, if we take

$$\begin{aligned} x &= (x_1, \ldots, x_{s+2t}) \\ x' &= (x_1', \ldots, x_{s+2t}') \end{aligned}$$

with respect to the *new* co-ordinates (4), then the inner product is defined by

$$(x, x') = x_1 x_1' + \ldots + x_{s+2t} x_{s+2t}'.$$

The *length* of a vector x is then

$$\|x\| = \sqrt{(x, x)}$$

and the *distance* between x and x' is $\|x - x'\|$.

Referred to our original mixture of real and complex

co-ordinates we have, for

$$x = (x_1, \ldots, x_s; y_1 + iz_1, \ldots, y_t + iz_t),$$

$$\|x\| = \sqrt{(x_1^2 + \ldots + x_s^2 + y_1^2 + z_1^2 + \ldots + y_t^2 + z_t^2)}.$$

Exercises

8.1 Find the monomorphisms $\sigma_i : K \to \mathbf{C}$ for the following fields and determine the number s of the σ_i satisfying $\sigma_i(K) \subseteq \mathbf{R}$, and the number t of distinct conjugate pairs σ_i, σ_j such that $\bar{\sigma}_i = \sigma_j$:
(i) $\mathbf{Q}(\sqrt{5})$ (ii) $\mathbf{Q}(\sqrt{-5})$ (iii) $\mathbf{Q}(\sqrt[4]{5})$
(iv) $\mathbf{Q}(\zeta)$ where $\zeta = e^{2\pi i/7}$
(v) $\mathbf{Q}(\zeta)$ where $\zeta = e^{2\pi i/p}$ for a rational prime p.

8.2 For $K = \mathbf{Q}(\sqrt{d})$ where d is a squarefree integer, calculate $\sigma : K \to \mathbf{L}^{st}$, distinguishing the cases $d < 0$, $d > 0$. Compute $\mathrm{N}(x)$ for $x \in \mathbf{L}^{st}$ and by direct calculation verify

$$\mathrm{N}(\alpha) = \mathrm{N}(\sigma(\alpha)) \quad (\alpha \in K).$$

8.3 Let $K = \mathbf{Q}(\theta)$ where the algebraic integer θ has minimum polynomial f. If f factorizes over $\mathbf{R}$ into irreducibles as

$$f(t) = g_1(t) \ldots g_q(t) h_1(t) \ldots h_r(t)$$

where g_i is linear and h_j quadratic, prove that $q = s$ and $r = t$ in the notation of the chapter for s, t.

8.4 Let $K = \mathbf{Q}(\theta)$ where $\theta \in \mathbf{R}$ and $\theta^3 = 3$. What is the map σ in this case? Pick a basis for K and verify Theorem 8.1 for it.

8.5 Find a map from $\mathbf{R}^2$ to itself under which $\mathbf{Q}$-linearly independent sets map to $\mathbf{Q}$-linearly independent sets, but some $\mathbf{R}$-linearly independent set does not map to an $\mathbf{R}$-linearly independent set.

8.6 If $K = \mathbf{Q}(\theta)$ where $\theta \in \mathbf{R}$ and $\theta^3 = 3$, verify Corollary 8.3 for the additive subgroup of K generated by $1 + \theta$ and $\theta^2 - 2$.

CHAPTER NINE

Class-group and class-number

The class-group of a number field is the quotient of the group of fractional ideals by the (normal) subgroup of principal fractional ideals; the class-number is its order. In some sense the class-group measures the extent to which ideals can be non-principal, or the extent to which factorization fails to be unique. In particular the factorization of elements of $\mathfrak{O}$ is unique if and only if the class-number is 1.

Minkowski's theorem is used to prove finiteness of the class-number. Simple group-theoretic considerations yield useful conditions for an ideal to be principal. These lead to a proof that every ideal becomes principal in a suitable extension field, which is one formulation of the basic idea of Kummer's 'ideal numbers' within the ideal-theoretic framework.

The importance of the class-number can only be hinted at here. It will be crucial to our investigation of Fermat's conjecture in Chapter 11. Many deep and delicate results in the theory of numbers are related to arithmetical properties of the class-number or algebraic properties of the class-group.

9.1 The class-group

As usual let $\mathfrak{O}$ be the ring of integers of a number field K of degree n. Theorem 5.15 tells us that prime factorization in $\mathfrak{O}$ is unique if and only if every ideal of $\mathfrak{O}$ is principal. Our aim here is to find a way of measuring how far prime factorization fails to be unique in the case where $\mathfrak{O}$ contains non-principal ideals, or equivalently how far away the ideals of $\mathfrak{O}$ are from being principal.

To this end we use the group of fractional ideals defined in Chapter 5. Say that a fractional ideal of $\mathfrak{O}$ is *principal* if it is of the form $c^{-1}\mathfrak{a}$ where $\mathfrak{a}$ is a principal ideal of $\mathfrak{O}$. Let $\mathscr{F}$ be the group of fractional ideals under multiplication. It is easy to check that the set $\mathscr{P}$ of principal fractional ideals is a subgroup of $\mathscr{F}$. We define the *class-group* of $\mathfrak{O}$ to be the quotient group

$$\mathscr{H} = \mathscr{F}/\mathscr{P}.$$

The *class-number* $h = h(\mathfrak{O})$ is defined to be the order of $\mathscr{H}$.

Since each of $\mathscr{F}$, $\mathscr{P}$ is an infinite group we have no immediate way of deciding whether or not h is finite. In fact it is; and our subsequent efforts will be devoted to a proof of this deep and important fact. First, however, we shall reformulate the definition of the class-group in a manner independent of fractional ideals.

Let us say that two fractional ideals are *equivalent* if they belong to the same coset of $\mathscr{P}$ in $\mathscr{F}$, or in other words if they map to the same element of $\mathscr{F}/\mathscr{P}$. If $\mathfrak{a}$ and $\mathfrak{b}$ are fractional ideals we write

$$\mathfrak{a} \sim \mathfrak{b}$$

if $\mathfrak{a}$ and $\mathfrak{b}$ are equivalent, and use

$$[\mathfrak{a}]$$

to denote the equivalence class of $\mathfrak{a}$.

The class-group $\mathscr{H}$ is the set of these equivalence classes.

If $\mathfrak{a}$ is a fractional ideal then $\mathfrak{a} = c^{-1}\mathfrak{b}$ where $c \in \mathfrak{O}$ and $\mathfrak{b}$ is an ideal. Hence

$$\mathfrak{b} = c\mathfrak{a} = \langle c\rangle\mathfrak{a}$$

and since $\langle c\rangle \in \mathscr{P}$ this means that $\mathfrak{a} \sim \mathfrak{b}$. In other words, *every equivalence class contains an ideal.*

Now let $\mathfrak{x}$ and $\mathfrak{y}$ be equivalent ideals. Then $\mathfrak{x} = \mathfrak{c}\mathfrak{y}$ where $\mathfrak{c}$ is a principal fractional ideal, say $\mathfrak{c} = d^{-1}\mathfrak{e}$ for $d \in \mathfrak{O}$, $\mathfrak{e}$ a principal ideal. Therefore

$$\mathfrak{x}\langle d \rangle = \mathfrak{y}\mathfrak{e} .$$

Conversely if $\mathfrak{x}\mathfrak{d} = \mathfrak{y}\mathfrak{e}$ for $\mathfrak{d}, \mathfrak{e}$ principal ideals, then $\mathfrak{x} \sim \mathfrak{y}$.

This allows us to describe $\mathscr{H}$ as follows: take the set $\mathscr{I}$ of all ideals, and define upon it a relation $\sim$ by $\mathfrak{x} \sim \mathfrak{y}$ if and only if there exist *principal* ideals $\mathfrak{d}, \mathfrak{e}$ with $\mathfrak{x}\mathfrak{d} = \mathfrak{y}\mathfrak{e}$. Then $\mathscr{H}$ is the set of equivalence classes $[\mathfrak{x}]$, with group operation defined by

$$[\mathfrak{x}][\mathfrak{y}] = [\mathfrak{x}\mathfrak{y}].$$

It is for this reason that $\mathscr{H}$ is called the *class*-group.

The significance of the class-group is that it captures the extent to which factorization is not unique. In particular we have

Theorem 9.1. *Factorization in* $\mathfrak{O}$ *is unique if and only if the class-group* $\mathscr{H}$ *has order* 1, *or equivalently the class-number* $h = 1$.

Proof. Factorization is unique if and only if every ideal of $\mathfrak{O}$ is principal (Theorem 5.15), which in turn is true if and only if every fractional ideal is principal, which is equivalent to $\mathscr{F} = \mathscr{P}$, which is equivalent to $|\mathscr{H}| = h = 1$. □

The rest of this chapter proves that h is finite, and deduces a few useful consequences. In the next chapter we develop some methods whereby h, and the structure of $\mathscr{H}$, may be computed: such methods are an obvious necessity for applications of the class-group in particular cases.

9.2 An existence theorem

The finiteness of h rests on an application of Minkowski's theorem to the space $\mathbf{L}^{st}$. It is, in fact, possible to give a more elementary proof that h is finite, see [6] for example, but

Minkowski's theorem gives a better bound, and is in any case needed elsewhere. In this section we state and prove the relevant result, leaving the finiteness theorem to the next section.

Lemma 9.2 *If M is a lattice in $\mathbf{L}^{st}$ of dimension $s + 2t$ having fundamental domain of volume V, and if $c_1, \ldots, c_{s+t}$ are positive real numbers whose product*

$$c_1 \ldots c_{s+t} > \left(\frac{4}{\pi}\right)^t V$$

then there exists in M a non-zero element

$$x = (x_1, \ldots, x_{s+t})$$

such that

$$|x_1| < c_1, \ldots, |x_s| < c_s;$$

$$|x_{s+1}|^2 < c_{s+1}, \ldots, |x_{s+t}|^2 < c_{s+t}.$$

Proof. Let X be the set of all points $x \in \mathbf{L}^{st}$ for which the conclusion holds. We compute

$$\begin{aligned} v(X) &= \int_{-c_1}^{c_1} \mathrm{d}x_1 \ldots \int_{-c_s}^{c_s} \mathrm{d}x_s \\ &\quad \times \iint_{y_1^2+z_1^2<c_{s+1}} \mathrm{d}y_1\,\mathrm{d}z_1 \ldots \\ &\quad \times \iint_{y_t^2+z_t^2<c_{s+t}} \mathrm{d}y_t\,\mathrm{d}z_t \\ &= 2c_1 . 2c_2 \ldots 2c_s . \pi c_{s+1} \ldots \pi c_{s+t} \\ &= 2^s \pi^t c_1 \ldots c_{s+t}. \end{aligned}$$

Now X is a cartesian product of line segments and circular discs, so X is bounded, symmetric, and convex. Minkowski's theorem yields the required result provided

$$2^s \pi^t c_1 \ldots c_{s+t} > 2^{s+2t} V,$$

that is

$$c_1 \ldots c_{s+t} > \left(\frac{4}{\pi}\right)^t V. \qquad \square$$

Let K be a number field of degree $n = s + 2t$ as usual, with ring of integers $\mathfrak{O}$; and let $\mathfrak{a}$ be an ideal of $\mathfrak{O}$. Then $(\mathfrak{a}, +)$ is a free abelian group of rank n (Theorem 2.15) so by Corollary 8.3 its image $\sigma(\mathfrak{a})$ in $\mathbf{L}^{st}$ is a lattice of dimension n. To use the above lemma in this situation we must know the volume of a fundamental domain for $\sigma(\mathfrak{a})$. First we note:

Lemma 9.3. *Let L be an n-dimensional lattice in $\mathbf{R}^n$ with basis $\{e_1, \ldots, e_n\}$. Suppose*

$$e_i = (a_{1i}, \ldots, a_{ni}).$$

Then the volume of the fundamental domain T of L defined by this basis is

$$v(T) = |\det a_{ij}|.$$

Proof. We have

$$v(T) = \int_T \mathrm{d}x_1 \ldots \mathrm{d}x_n.$$

Define new variables by

$$x_i = \sum_j a_{ij} y_j.$$

The Jacobian of this transformation is equal to $\det a_{ij}$, and T is the set of points $\Sigma\, b_i y_i$ with $0 \leqslant b_i < 1$. By the transformation formula for multiple integrals (Apostol [15] p. 271) we have

$$\begin{aligned} v(T) &= \int_T |\det a_{ij}|\, \mathrm{d}y_1 \ldots \mathrm{d}y_n \\ &= |\det a_{ij}| \int_0^1 \mathrm{d}y_1 \ldots \int_0^1 \mathrm{d}y_n \\ &= |\det a_{ij}|. \end{aligned} \qquad \square$$

Given a lattice L there exist many different $\mathbf{Z}$-bases for L, and hence many distinct fundamental domains. However, since distinct **Z**-bases are related by a unimodular matrix, it follows from Lemma 9.3 that the volumes of these distinct fundamental domains are all equal.

Theorem 9.4. *Let K be a number field of degree $n = s + 2t$ as usual, with ring of integers $\mathfrak{O}$, and let $0 \neq \mathfrak{a}$ be an ideal of $\mathfrak{O}$. Then the volume of a fundamental domain for $\sigma(\mathfrak{a})$ in $\mathbf{L}^{st}$ is equal to*

$$2^{-t}N(\mathfrak{a})\sqrt{|\Delta|}$$

where Δ is the discriminant of K.

Proof. Let $\{\alpha_1, \ldots, \alpha_n\}$ be a **Z**-basis for $\mathfrak{a}$. Then, in the notation of Theorem 8.1, a **Z**-basis for $\sigma(\mathfrak{a})$ in $\mathbf{L}^{st}$ is

$$(x_1^{(1)}, \ldots, x_s^{(1)}, y_1^{(1)}, z_1^{(1)}, \ldots, y_t^{(1)}, z_t^{(1)}),$$

$$\cdots\cdots\cdots\cdot$$

$$(x_1^{(n)}, \ldots, x_s^{(n)}, y_1^{(n)}, z_1^{(n)}, \ldots, y_t^{(n)}, z_t^{(n)}).$$

Hence by Lemma 9.3, if T is a fundamental domain for $\sigma(\mathfrak{a})$, we have

$$v(T) = |D|,$$

where D is as in Theorem 8.1. Using the notation of that theorem we have

$$D = (-2i)^{-t}E$$

so that

$$|D| = 2^{-t}|E|.$$

Now $E^2 = \Delta[\alpha_1, \ldots, \alpha_n]$ and

$$N(\mathfrak{a}) = \left|\frac{\Delta[\alpha_1, \ldots, \alpha_n]}{\Delta}\right|^{1/2}$$

by Theorem 5.8, whence the result. □

We may now apply Theorem 9.4 and Lemma 9.2 to yield the important:

Theorem 9.5. *If* $\mathfrak{a} \neq 0$ *is an ideal of* $\mathfrak{O}$ *then* $\mathfrak{a}$ *contains an integer* α *with*

$$|N(\alpha)| \leqslant \left(\frac{2}{\pi}\right)^t N(\mathfrak{a})\sqrt{|\Delta|}$$

where Δ *is the discriminant of K.*

Proof. For a fixed but arbitrary $\epsilon > 0$ choose positive real numbers $c_1, \ldots, c_{s+t}$ with

$$c_1 \ldots c_{s+t} = \left(\frac{2}{\pi}\right)^t N(\mathfrak{a})\sqrt{|\Delta|} + \epsilon.$$

By Lemma 9.2 and Theorem 9.4 it follows that there exists $0 \neq \alpha \in \mathfrak{a}$ such that

$$|\sigma_1(\alpha)| < c_1, \ldots, |\sigma_s(\alpha)| < c_s,$$
$$|\sigma_{s+1}(\alpha)|^2 < c_{s+1}, \ldots, |\sigma_{s+t}(\alpha)|^2 < c_{s+t}.$$

Multiplying all these inequalities together we obtain

$$|N(\alpha)| < c_1 \ldots c_s c_{s+1} \ldots c_{s+t}$$
$$= \left(\frac{2}{\pi}\right)^t N(\mathfrak{a})\sqrt{|\Delta|} + \epsilon.$$

Since a lattice is discrete, it follows that the set A_ϵ of such α is finite. Also $A_\epsilon \neq \emptyset$, so that $A = \cap_\epsilon A_\epsilon \neq \emptyset$. If we pick $\alpha \in A$ then

$$|N(\alpha)| \leqslant \left(\frac{2}{\pi}\right)^t N(\mathfrak{a})\sqrt{|\Delta|}. \qquad \square$$

Corollary 9.6. *Every non-zero ideal* $\mathfrak{a}$ *of* $\mathfrak{O}$ *is equivalent to an ideal whose norm is* $\leqslant (2/\pi)^t\sqrt{|\Delta|}$.

Proof. The class of fractional ideals equivalent to $\mathfrak{a}^{-1}$ contains an ideal $\mathfrak{c}$, so $\mathfrak{a}\mathfrak{c} \sim \mathfrak{O}$. We can use Theorem 9.5 to find an integer $\gamma \in \mathfrak{c}$ such that

$$|N(\gamma)| \leqslant \left(\frac{2}{\pi}\right)^t N(\mathfrak{c})\sqrt{|\Delta|}.$$

Since $\mathfrak{c} \mid \gamma$ we have

$$\langle \gamma \rangle = \mathfrak{c}\mathfrak{b}$$

for some ideal $\mathfrak{b}$. Since $N(\mathfrak{b})N(\mathfrak{c}) = N(\mathfrak{b}\mathfrak{c}) = N(\langle \gamma \rangle) = |N(\gamma)|$ we have

$$N(\mathfrak{b}) \leqslant \left(\frac{2}{\pi}\right)^t \sqrt{|\Delta|}.$$

We claim that $\mathfrak{b} \sim \mathfrak{a}$. This is clear since $\mathfrak{c} \sim \mathfrak{a}^{-1}$ and $\mathfrak{b} \sim \mathfrak{c}^{-1}$. □

An explicit computation. If $K = \mathbf{Q}(\sqrt{-5})$, then $\mathfrak{O} = \mathbf{Z}[\sqrt{-5}]$ does not have unique factorization, so $h > 1$. Because the monomorphisms $\sigma_i : K \to \mathbf{C}$ are σ_1, σ_2 where $\sigma_1 \neq \sigma_2$ and $\bar{\sigma}_1 = \sigma_2$, we have $t = 1$. The discriminant Δ of K is $\Delta = -20$. Hence

$$\left(\frac{2}{\pi}\right)^t \sqrt{|\Delta|} = \frac{2\sqrt{20}}{\pi} < 2.85.$$

Every ideal of $\mathfrak{O}$ is then equivalent to an ideal of norm less than 2.85, which means a norm of 1 or 2. An ideal of norm 1 is the whole ring $\mathfrak{O}$, hence principal. An ideal $\mathfrak{a}$ of norm 2 satisfies $\mathfrak{a} \mid 2$ by Theorem 5.11(b), so $\mathfrak{a}$ is a factor of $\langle 2 \rangle$. But

$$\langle 2 \rangle = \langle 2, 1 + \sqrt{-5} \rangle^2$$

where $\langle 2, 1 + \sqrt{-5} \rangle$ is prime and has norm 2. So $\langle 2, 1 + \sqrt{-5} \rangle$ is the *only* ideal of norm 2. Hence every ideal of $\mathfrak{O}$ is equivalent to $\mathfrak{O}$ or $\langle 2, 1 + \sqrt{-5} \rangle$ which are themselves inequivalent (since $\langle 2, 1 + \sqrt{-5} \rangle$ is not principal), proving $h = 2$.

9.3 Finiteness of the class-group

Theorem 9.7. *The class-group of a number field is a finite abelian group. The class-number h is finite.*

Proof. Let K be a number field, of discrimanant Δ, and degree $n = s + 2t$ as usual. We know that the class-group $\mathscr{H} = \mathscr{F}/\mathscr{P}$ is abelian, so it remains to prove $\mathscr{H}$ finite: this is true if and only if the number of distinct equivalence classes of fractional ideals is finite. Let $[\mathfrak{c}]$ be such an equivalence class. Then $[\mathfrak{c}]$ contains an ideal $\mathfrak{a}$, and by Corollary 9.6. $\mathfrak{a}$ is equivalent to an ideal $\mathfrak{b}$ with $N(\mathfrak{b}) \leqslant (2/\pi)^t\sqrt{|\Delta|}$. Since only finitely many ideals have a given norm (Theorem 5.12c) there are only finitely many choices for $\mathfrak{b}$. Since $[\mathfrak{c}] = [\mathfrak{a}] = [\mathfrak{b}]$ (because $\mathfrak{c} \sim \mathfrak{a} \sim \mathfrak{b}$) it follows that there are only finitely many equivalence classes $[\mathfrak{c}]$, whence $\mathscr{H}$ is a finite group and $h = |\mathscr{H}|$ is finite. □

From simple group-theoretic facts we obtain the useful:

Proposition 9.8. *Let K be a number field of class-number h, and* $\mathfrak{a}$ *an ideal of the ring of integers* $\mathfrak{O}$. *Then*

(a) $\mathfrak{a}^h$ *is principal,*

(b) *If q is prime to h and* $\mathfrak{a}^q$ *is principal, then* $\mathfrak{a}$ *is principal.*

Proof. Since $h = |\mathscr{H}|$ we have $[\mathfrak{a}]^h = [\mathfrak{O}]$ for all $[\mathfrak{a}] \in \mathscr{H}$, because $[\mathfrak{O}]$ is the identity element of $\mathscr{H}$. Hence $[\mathfrak{a}^h] = [\mathfrak{a}]^h = [\mathfrak{O}]$, so $\mathfrak{a}^h \sim \mathfrak{O}$, so $\mathfrak{a}^h$ is principal. This proves (a). For (b) choose u and $v \in \mathbf{Z}$ such that $uh + vq = 1$. Then $[\mathfrak{a}]^q = [\mathfrak{O}]$, so we have

$$\begin{aligned}[\mathfrak{a}] &= [\mathfrak{a}]^{uh+vq}\\ &= ([\mathfrak{a}]^h)^u([\mathfrak{a}]^q)^v\\ &= [\mathfrak{O}]^u[\mathfrak{O}]^v\\ &= [\mathfrak{O}]\end{aligned}$$

hence again $\mathfrak{a}$ is principal. □

9.4 How to make an ideal principal

This section and the next are not required elsewhere in the book and may be omitted if so desired.

Given an ideal $\mathfrak{a}$ in the ring $\mathfrak{O}$ of integers of a number field K, we already know that $\mathfrak{a}$ has at most two generators

$$\mathfrak{a} = \langle \alpha, \beta \rangle \qquad (\alpha, \beta \in \mathfrak{O}).$$

What we demonstrate in this section is that we can find an extension number field $E \supseteq K$ with integers $\mathfrak{O}'$ such that the extended ideal $\mathfrak{O}'\mathfrak{a}$ in $\mathfrak{O}'$ is principal. As a standard piece of notation we shall retain the symbols $\langle\alpha\rangle$, $\langle\alpha, \beta\rangle$ to denote the ideals in $\mathfrak{O}$ generated by α and by $\alpha, \beta \in \mathfrak{O}$, whilst writing the ideal in $\mathfrak{O}'$ generated by $S \subseteq \mathfrak{O}'$ as $\mathfrak{O}'S$. For example $\mathfrak{O}'\kappa$ will denote the principal ideal in $\mathfrak{O}'$ generated by $\kappa \in \mathfrak{O}'$.

Lemma 9.9. *If S_1, S_2 are subsets of $\mathfrak{O}'$, then*

$$\mathfrak{O}'(S_1S_2) = (\mathfrak{O}'S_1)(\mathfrak{O}'S_2).$$

Proof. Trivial (remembering $1 \in \mathfrak{O}'$). □

The central result is then:

Theorem 9.10. *Let K be a number field, $\mathfrak{a}$ an ideal in the ring of integers $\mathfrak{O}$ of K. Then there exists an algebraic integer κ such that if $\mathfrak{O}'$ is the ring of integers of $K(\kappa)$, then*

(i) $\mathfrak{O}'\kappa = \mathfrak{O}'\mathfrak{a}$
(ii) $(\mathfrak{O}'\kappa) \cap \mathfrak{O} = \mathfrak{a}$
(iii) *If* $\mathbf{B}$ *is the ring of all algebraic integers, then* $(\mathbf{B}\kappa) \cap K = \mathfrak{a}$.
(iv) *If* $\mathfrak{O}''\gamma = \mathfrak{O}''\mathfrak{a}$ *for any* $\gamma \in \mathbf{B}$, *and any ring* $\mathfrak{O}''$ *of integers, then* $\gamma = u\kappa$ *where* u *is a unit of* $\mathbf{B}$.

Proof. By proposition 9.8, $\mathfrak{a}^h$ is principal, say $\mathfrak{a}^h = \langle\omega\rangle$. Let $\kappa = \omega^{1/h} \in \mathbf{B}$, and consider $E = K(\kappa)$. Let $\mathfrak{O}' = \mathbf{B} \cap E$ be the ring of integers in E, then clearly $\kappa \in \mathfrak{O}'$. Since $\mathfrak{a}^h = \langle\omega\rangle$, it follows, using Lemma 9.9, that

$$(\mathfrak{O}'\mathfrak{a})^h = \mathfrak{O}'(\mathfrak{a}^h) = \mathfrak{O}'\omega = \mathfrak{O}'\kappa^h = (\mathfrak{O}'\kappa)^h.$$

Uniqueness of factorization of ideals in $\mathfrak{O}'$ easily yields

$$\mathfrak{O}'\mathfrak{a} = \mathfrak{O}'\kappa,$$

proving (i)

Since (iii) implies (ii), we now consider (iii). The inclusion $\mathfrak{a} \subseteq \mathbf{B}\kappa \cap K$ is straightforward. Conversely, suppose $\gamma \in \mathbf{B}\kappa \cap K$, then

$$\gamma = \lambda\kappa \qquad (\lambda \in \mathbf{B})$$

and we are required to show $\gamma \in \mathfrak{a}$. First note that, since $\gamma \in K$, $\kappa \in E$, we have $\lambda = \gamma\kappa^{-1} \in E$, so $\lambda \in E \cap \mathbf{B} = \mathfrak{O}'$.

This gives

$$\gamma^h = \lambda^h\kappa^h = \lambda^h\omega \qquad (\gamma \in K, \lambda \in \mathfrak{O}', \omega \in \mathfrak{O})$$

so $\gamma^h \in \mathbf{B}$, and by Theorem 2.9, $\gamma \in \mathbf{B}$. Thus $\gamma \in \mathbf{B} \cap K = \mathfrak{O}$. Considering the equation $\gamma^h = \lambda^h\omega$ again, we find

$$\lambda^h = \gamma^h\omega^{-1} \in K$$

so $\lambda^h \in K \cap \mathbf{B} = \mathfrak{O}$. Thus we finish up with

$$\gamma^h = \lambda^h\omega \qquad (\gamma, \lambda^h, \omega \in \mathfrak{O}).$$

Taking ideals in $\mathfrak{O}$ we get

$$\langle\gamma\rangle^h = \langle\lambda^h\rangle\langle\omega\rangle = \langle\lambda^h\rangle\mathfrak{a}^h.$$

Unique factorization in $\mathfrak{O}$ implies $\langle\lambda^h\rangle = \mathfrak{b}^h$ for some ideal $\mathfrak{b}$, so

$$\langle\gamma\rangle^h = \mathfrak{b}^h\mathfrak{a}^h$$

and unique factorization once more implies

$$\langle\gamma\rangle = \mathfrak{b}\mathfrak{a},$$

whence $\gamma \in \mathfrak{a}$, as required.

The proof of (iv) is found by noting that by Theorem 5.14, $\mathfrak{a} = \langle\alpha, \beta\rangle$ for $\alpha, \beta \in \mathfrak{O}$; and substituting in (iv) gives

$$\mathfrak{O}''\gamma = \mathfrak{O}''\langle\alpha, \beta\rangle.$$

Thus

$$\gamma = \lambda\alpha + \mu\beta$$

where $\lambda, \mu \in \mathfrak{O}''$, so certainly $\lambda, \mu \in \mathbf{B}$. From (i), $\alpha, \beta \in \mathfrak{O}'\kappa$, so

$$\alpha = \eta\kappa, \qquad \beta = \zeta\kappa \qquad (\eta, \zeta \in \mathfrak{O}' \subseteq \mathbf{B}).$$

Hence $\gamma = \lambda\eta\kappa + \mu\zeta\kappa$ and $\kappa \mid \gamma$ in $\mathbf{B}$. Interchanging the roles of γ, κ proves (iv). □

Theorem 9.10 can be improved, for as it stands the extension ring $\mathfrak{O}'$ in which $\mathfrak{O}'\mathfrak{a}$ is principal depends on $\mathfrak{a}$. We can actually find a single extension ring in which the extension of every ideal is principal. This depends on the following Lemma and the finiteness of the class-number:

Lemma 9.11. *If $\mathfrak{a}, \mathfrak{b}$ are equivalent ideals in the ring $\mathfrak{O}$ of integers of a number field and $\mathfrak{O}'\mathfrak{a}$ is principal, then so is $\mathfrak{O}'\mathfrak{b}$.*

Proof. By the definition of equivalence, there exist principal ideals $\mathfrak{d}, \mathfrak{e}$ of $\mathfrak{O}$ such that $\mathfrak{a}\mathfrak{d} = \mathfrak{b}\mathfrak{e}$. Hence

$$(\mathfrak{O}'\mathfrak{a})(\mathfrak{O}'\mathfrak{d}) = (\mathfrak{O}'\mathfrak{b})(\mathfrak{O}'\mathfrak{e})$$

where now $\mathfrak{O}'\mathfrak{a}$, $\mathfrak{O}'\mathfrak{d}$, $\mathfrak{O}'\mathfrak{e}$ are all principal. Since the set $\mathscr{P}$ of principal fractional ideals of $\mathfrak{O}'$ is a group, $\mathfrak{O}'\mathfrak{b}$ is a principal fractional ideal which is also an ideal, so $\mathfrak{O}'\mathfrak{b}$ is a principal ideal. □

Theorem 9.12. *Let K be a number field with integers $\mathfrak{O}_K$, then there exists a number field $L \supseteq K$ with ring $\mathfrak{O}_L$ of integers such that for every ideal $\mathfrak{a}$ in $\mathfrak{O}_K$ we have*
(i) *$\mathfrak{O}_L\mathfrak{a}$ is a principal ideal,*
(ii) *$(\mathfrak{O}_L\mathfrak{a}) \cap \mathfrak{O}_K = \mathfrak{a}$.*

Proof. Since h is finite, select a representative set of ideals $\mathfrak{a}_1, \ldots, \mathfrak{a}_h$, one from each class and choose algebraic integers $\kappa_1, \ldots, \kappa_h$ such that $\mathfrak{O}_i\mathfrak{a}_i$ is principal where $\mathfrak{O}_i$ is the ring of integers of $K(\kappa_i)$. Let $L = K(\kappa_1, \ldots, \kappa_h)$, then its ring $\mathfrak{O}_L$ of integers contains all the $\mathfrak{O}_i$. Hence each ideal $\mathfrak{O}_L\mathfrak{a}_i$ is principal in $\mathfrak{O}_L$. Since every ideal $\mathfrak{a}$ in $\mathfrak{O}$ is equivalent to some $\mathfrak{a}_i$, then $\mathfrak{O}_L\mathfrak{a}$ is principal by Lemma 9.11, say

$$\mathfrak{O}_L\mathfrak{a} = \mathfrak{O}_L\alpha \qquad (\alpha \in \mathbf{B}).$$

This proves (i)

Clearly $\mathfrak{a} \subseteq (\mathfrak{O}_L\mathfrak{a}) \cap \mathfrak{O}_K$. For the converse inclusion, Theorem 9.10 (iv) implies $\alpha = u\kappa$ where u is a unit in $\mathbf{B}$. Hence

$$\begin{aligned}(\mathfrak{O}_L\mathfrak{a}) \cap \mathfrak{O}_K &= (\mathfrak{O}_L\alpha) \cap \mathfrak{O}_K \\ &\subseteq (\mathbf{B}\alpha) \cap K \\ &= (\mathbf{B}\kappa) \cap K \\ &= \mathfrak{a}\end{aligned}$$

by Theorem 9.10 (iii). □

Until quite recently it was an open question, going back to Hilbert, whether every number field can be embedded in one with unique factorization. However, in 1964 Golod and Šafarevič [26] showed that this was not always possible, citing the explicit example of $\mathbf{Q}(\sqrt{(-3.5.7.11.13.17.19)})$. The proof is ingenious rather than hard, but uses ideas we have not developed.

9.5 Unique factorization of elements in an extension ring

The results of the last section can be translated from principal ideals back to elements to give the version of Kummer's theory alluded to in the introduction to Chapter 5. There we considered examples of non-unique factorization such as

$$10 = 2.5 = (5+\sqrt{15})(5-\sqrt{15})$$

in the ring of integers of $\mathbf{Q}(\sqrt{15})$. Viewing this as an equation in $\mathbf{Q}(\sqrt{3},\sqrt{5})$, we saw that the factors could be further reduced as

$$\begin{aligned} 2 &= (\sqrt{5}+\sqrt{3})(\sqrt{5}-\sqrt{3}) \\ 5 &= \sqrt{5}\sqrt{5}, \\ 5+\sqrt{15} &= \sqrt{5}(\sqrt{5}+\sqrt{3}) \\ 5-\sqrt{15} &= \sqrt{5}(\sqrt{5}-\sqrt{3}) \end{aligned}$$

and the two factorizations of 10 found above were just regroupings of the factors

$$10 = \sqrt{5}\sqrt{5}(\sqrt{5}+\sqrt{3})(\sqrt{5}-\sqrt{3}).$$

We can now show that such a phenomenon occurs in all cases in rings of integers.

Theorem 9.13. *Suppose K is a number field with integers $\mathfrak{O}_K$. Then there exists an extension field $L \supseteq K$ with integers $\mathfrak{O}_L$ such that every non-zero, non-unit $a \in \mathfrak{O}_K$ has a factorization*

$$a = p_1 \ldots p_r \qquad (p_i \in \mathfrak{O}_L)$$

where the p_i are non-units in $\mathfrak{O}_L$, and the following property is satisfied:

Given any factorization in $\mathfrak{O}_K$*:*

$$a = a_1 \ldots a_s$$

where the a_i *are non-units in* $\mathfrak{O}_K$*, there exist integers*

$$1 \leqslant n_1 < \ldots < n_s = r$$

and a permutation π *of* $\{1, \ldots, r\}$ *such that the following elements are associates in* $\mathfrak{O}_L$*:*

$$a_1, p_{\pi(1)} \ldots p_{\pi(n_1)}$$
$$\ldots\ldots\ldots\ldots$$
$$a_s, p_{\pi(n_{s-1}+1)} \ldots p_{\pi(n_s)}.$$

Remark. What this theorem says in plain language is that the factorizations of elements into irreducibles in $\mathfrak{O}_K$ may not be unique, but all factorizations of an element in $\mathfrak{O}_K$ come from different groupings of associates of a single factorization in $\mathfrak{O}_L$. In this sense elements in $\mathfrak{O}_K$ have unique factorization into elements in $\mathfrak{O}_L$.

Proof. There is a unique factorization of $\langle a\rangle$ into prime ideals in $\mathfrak{O}_K$, say

$$\langle a\rangle = \mathfrak{p}_1 \ldots \mathfrak{p}_r.$$

Since a is a non-unit, we have $r \geqslant 1$. Let $\mathfrak{O}_L$ be a ring of integers as in Theorem 9.12 where every ideal of $\mathfrak{O}_K$ extends to a principal ideal, and suppose

$$\mathfrak{O}_L \mathfrak{p}_i = \mathfrak{O}_L p_i \qquad (p_i \in \mathfrak{O}_L).$$

Then $a = up_1 \ldots p_r$ where u is a unit in $\mathfrak{O}_L$, and since $r \geqslant 1$, we may replace p_1 by $up_1 \in \mathfrak{O}_L p_1$ to get a factorization of the form

$$a = p_1 \ldots p_r.$$

Given any factorization of elements

$$a = a_1 \ldots a_s$$

where the a_i are non-units in $\mathfrak{O}_K$, then

$$\langle a\rangle = \langle a_1\rangle \ldots \langle a_s\rangle,$$

where all the $\langle a_i \rangle$ are proper ideals. Unique factorization in $\mathfrak{O}_K$ gives us the integers n_i and the permutation π such that

$$\langle a_1 \rangle = \mathfrak{p}_{\pi(1)} \cdots \mathfrak{p}_{\pi(n_1)}$$

$$\cdots\cdots\cdots\cdots$$

$$\langle a_s \rangle = \mathfrak{p}_{\pi(n_{s-1}+1)} \cdots \mathfrak{p}_{\pi(n_s)}.$$

Now take ideals in $\mathfrak{O}_L$ generated by these ideals and the result follows. □

Example. From the explicit computation of Section 2, if $K = \mathbf{Q}(\sqrt{-5})$, then $h = 2$ and a representative set of ideals are $\mathfrak{O}$, and $\langle 2, 1+\sqrt{-5}\rangle$ where $\langle 2, 1+\sqrt{-5}\rangle^2 = \langle 2 \rangle$. Hence we may take $L = K(\sqrt{2}) = \mathbf{Q}(\sqrt{-5}, \sqrt{2})$. Theorem 9.13 tells us that *every* element of $\mathbf{Z}[\sqrt{-5}]$ factorizes uniquely in the integers of $\mathbf{Q}(\sqrt{-5}, \sqrt{2})$. The case of the factorization of the element 6 will be dealt with in Exercise 9.7 where we shall find

$$6 = \sqrt{2}\sqrt{2}(\tfrac{1}{2}\sqrt{2} + \tfrac{1}{2}\sqrt{-10})\,(\tfrac{1}{2}\sqrt{2} - \tfrac{1}{2}\sqrt{-10}).$$

The fact that $\frac{1}{2}\sqrt{2} \pm \frac{1}{2}\sqrt{-10}$ really are integers may be dealt with by computing the explicit minimum polynomials of these elements over $\mathbf{Q}$. Granted this, it is an easy matter to check that the two alternative factorizations in $\mathbf{Z}[\sqrt{-5}]$:

$$6 = 2.3 = (1+\sqrt{-5})\,(1-\sqrt{-5})$$

are just different groupings of the factors in the integers of $\mathbf{Q}(\sqrt{-5}, \sqrt{2})$. (Do it!)

The above example shows up a basic problem in factorizing elements in an extension ring. We have not given a general method of computing the integers in a number field. To date we have only explicitly calculated the integers in quadratic and cyclotomic fields and those calculations were not trivial. There is also another weakness of factorizing elements in an extension ring. The elements p_i occurring in the factorization of a in Theorem 9.13 need not be irreducible. (For instance we might work in a slightly larger ring $\mathfrak{O}_{L'}$ containing $\sqrt{p_i}$; the method of adjoining $\kappa = \omega^{1/h}$ may very well add such roots.) However, the proof of Theorem 9.13 tells us that the factorization of the element a in $\mathfrak{O}_L$ which gives the unique factorization properties is given by the factorization of the ideal $\langle a \rangle$ in the ring $\mathfrak{O}_K$. For this reason we may just as well stick to ideals in the original ring rather than embellish the situation by factorizing elements outside. Our computations in future will be concerned mainly with ideals – unique factorization of ideals proves so much easier to handle!

Exercises

9.1 Let $K = \mathbf{Q}(\sqrt{-5})$, and let $\mathfrak{p}$, $\mathfrak{q}$, $\mathfrak{r}$ be the ideals defined in Exercise 5.2. Let $\mathscr{H}$ be the class group. Show that in $\mathscr{H}$ we have

$$[\mathfrak{p}]^2 = [\mathfrak{O}], \qquad [\mathfrak{p}][\mathfrak{q}] = [\mathfrak{O}], \qquad [\mathfrak{p}][\mathfrak{r}] = [\mathfrak{O}],$$

and hence show that $\mathfrak{p}, \mathfrak{q}, \mathfrak{r}$ are equivalent.

9.2 Verify by explicit computation that $\mathfrak{p}$, $\mathfrak{q}$, $\mathfrak{r}$ are equivalent.

9.3 Using Corollary 9.6, show that for $K = \mathbf{Q}(\sqrt{-6})$ every ideal is equivalent to one of norm at most 3. Verify

$$\langle 2 \rangle = \langle 2, \sqrt{-6} \rangle^2,$$
$$\langle 3 \rangle = \langle 3, \sqrt{-6} \rangle^2,$$

and conclude that the only ideals of norm 2, 3 are $\langle 2, \sqrt{-6} \rangle$, $\langle 3, \sqrt{-6} \rangle$. Deduce $h \leqslant 3$ and using $\langle 2, \sqrt{-6} \rangle^2 = \langle 2 \rangle$, or otherwise, show $h = 2$.

9.4 Find principal ideals $\mathfrak{a}$, $\mathfrak{b}$ in $\mathbf{Z}[\sqrt{-6}]$ such that

$$\mathfrak{a}\langle 2, \sqrt{-6} \rangle = \mathfrak{b}\langle 3, \sqrt{-6} \rangle.$$

9.5 Find all squarefree integers d in $-10 < d < 10$ such that the class-number of $\mathbf{Q}(\sqrt{d})$ is 1. (Hint: look up a few theorems!)

9.6 Using methods similar to Exercise 9.3, calculate the class-number of $\mathbf{Q}(\sqrt{d})$ for d squarefree and $-10 \leqslant d \leqslant 10$.

9.7 Suppose $K = \mathbf{Q}(\sqrt{-5})$, $\mathfrak{p} = \langle 2, 1+\sqrt{-5} \rangle$. Let $\mathfrak{O}'$ be the ring of integers of $\mathbf{Q}(\sqrt{-5}, \sqrt{2})$. Show $\mathfrak{O}'\mathfrak{p} = \mathfrak{O}'\sqrt{2}$. Find explicit integers $a, b \in \mathfrak{O}'$ such that

$$2 = \sqrt{2}a, \qquad 1+\sqrt{-5} = \sqrt{2}b,$$

and verify that a, b *are* integers by computing the monic polynomials which they satisfy over $\mathbf{Q}$. Using the notation of Exercise 9.1, find $\kappa_1, \kappa_2 \in \mathfrak{O}'$ such that

$$\mathfrak{O}'\kappa_1 = \mathfrak{O}'\mathfrak{q}, \qquad \mathfrak{O}'\kappa_2 = \mathfrak{O}'\mathfrak{r}$$

and use the factorization $\langle 6 \rangle = \mathfrak{p}^2\mathfrak{q}\mathfrak{r}$ to factorize the element 6 in $\mathfrak{O}'$. Explain how this factorization relates to

$$6 = 2.3 = (1+\sqrt{-5})(1+\sqrt{-5})$$

in $\mathbf{Z}[\sqrt{-5}]$.

9.8 In $\mathbf{Z}[\sqrt{-10}]$ we have the factorizations into irreducibles

$$14 = 2.7 = (2+\sqrt{-10})(2-\sqrt{-10}).$$

Find an extension ring $\mathfrak{O}_L$ of $\mathbf{Z}[\sqrt{-10}]$ and a factorization of 14 in $\mathfrak{O}_L$ such that the given factorizations are found by different groupings of the factors.

9.9 Factorize $6 = 2.3 = (4+\sqrt{10})(4-\sqrt{10}) \in \mathbf{Z}[\sqrt{10}]$ in an extension ring to exhibit the given factors as different groupings of the new ones.

9.10 Relate the factorization

$$10 = \sqrt{5}\sqrt{5}(\sqrt{5}+\sqrt{3})(\sqrt{5}-\sqrt{3})$$

in the integers of $\mathbf{Q}(\sqrt{3}, \sqrt{5})$ to the factorization of $\langle 10 \rangle$ into prime ideals in the integers of $\mathbf{Q}(\sqrt{15})$.

Explain how this gives rise to the different factorizations

$$10 = 2.5 = (5+\sqrt{15})(5-\sqrt{15})$$

into irreducibles in the integers of $\mathbf{Q}(\sqrt{15})$.

PART III

Number-theoretic applications

CHAPTER TEN

Computational methods

The results of this chapter, although apparently diverse, all have a strong bearing on the question of practical computation of the class-number, within the limits of the techniques now at our command. In the first section we study a special case of how a rational prime breaks up into prime ideals in a number field. The second section supplements this by showing that the distinct classes of fractional ideals may be found from the prime ideals dividing a finite set of rational primes, this set being in some sense 'small' provided the degree of K and its discriminant are not too 'large'. Several specific cases are studied, especially quadratic fields: in particular we complete the list of fields $\mathbf{Q}(\sqrt{d})$ with negative d and with class-number 1 (although we do not prove our list complete).

10.1 Factorization of a rational prime

If p is a prime number in $\mathbf{Z}$, it is not generally true that $\langle p\rangle$ is a prime ideal in the ring of integers $\mathfrak{O}$ of a number field K. For instance, in $\mathbf{Q}(\sqrt{-1})$ we have the factorization

$$\langle 2\rangle = \langle 1+\sqrt{-1}\rangle^2.$$

It is obviously useful to be able to compute the prime factors of $\langle p \rangle$. In the case where the ring of integers is generated by a single element (which includes quadratic and cyclotomic fields), the following theorem of Dedekind is decisive.

Theorem 10.1. *Let K be a number field of degree n with ring of integers* $\mathfrak{O} = \mathbf{Z}[\theta]$ *generated by* $\theta \in \mathfrak{O}$. *Given a rational prime p, suppose the minimum polynomial f of* θ *over* $\mathbf{Q}$ *gives rise to the factorization into irreducibles over* $\mathbf{Z}_p$:

$$\bar{f} = \bar{f}_1^{e_1} \dots \bar{f}_r^{e_r}$$

where the bar denotes the natural map $\mathbf{Z}[t] \to \mathbf{Z}_p[t]$. *Then if* $f_i \in \mathbf{Z}[t]$ *is any polynomial mapping onto* $\bar{f}_i$, *the ideal*

$$\mathfrak{p}_i = \langle p \rangle + \langle f_i(\theta) \rangle$$

is prime and the prime factorization of $\langle p \rangle$ *in* $\mathfrak{O}$ *is*

$$\langle p \rangle = \mathfrak{p}_1^{e_1} \dots \mathfrak{p}_r^{e_r}.$$

Proof. Let θ_i be a root of $\bar{f}_i$ in $\mathbf{Z}_p[\theta_i] \cong \mathbf{Z}_p[t]/\langle \bar{f}_i \rangle$. There is a natural map $\nu_i : \mathbf{Z}[\theta] \to \mathbf{Z}_p[\theta_i]$ given by

$$\nu_i(p(\theta)) = \bar{p}(\theta_i).$$

The image of ν_i is $\mathbf{Z}_p[\theta_i]$, which is a field, so $\ker \nu_i$ is a prime ideal of $\mathbf{Z}[\theta] = \mathfrak{O}$. Clearly

$$\langle p \rangle + \langle f_i(\theta) \rangle \subseteq \ker \nu_i.$$

But if $g(\theta) \in \ker \nu_i$, then $\bar{g}(\theta_i) = 0$, so $\bar{g} = \bar{f}_i \bar{h}$ for some $\bar{h} \in \mathbf{Z}_p[t]$; this means that $g - f_i h \in \mathbf{Z}[t]$ has coefficients divisible by p. Thus

$$\begin{aligned} g(\theta) &= (g(\theta) - f_i(\theta)h(\theta)) + f_i(\theta)h(\theta) \\ &\in \langle p \rangle + \langle f_i(\theta) \rangle, \end{aligned}$$

showing

$$\ker \nu_i = \langle p \rangle + \langle f_i(\theta) \rangle.$$

Let

$$\mathfrak{p}_i = \langle p \rangle + \langle f_i(\theta) \rangle,$$

then for each $\bar{f}_i$ the ideal $\mathfrak{p}_i$ is prime and satisfies $\langle p \rangle \subseteq \mathfrak{p}_i$ i.e. $\mathfrak{p}_i \mid \langle p \rangle$.

For any ideals $\mathfrak{a}$, $\mathfrak{b}_1$, $\mathfrak{b}_2$ we have

$$(\mathfrak{a} + \mathfrak{b}_1)(\mathfrak{a} + \mathfrak{b}_2) \subseteq \mathfrak{a} + \mathfrak{b}_1 \mathfrak{b}_2,$$

so by induction

$$\begin{aligned} \mathfrak{p}_1^{e_1} \ldots \mathfrak{p}_r^{e_r} &\subseteq \langle p \rangle + \langle f_1(\theta)^{e_1} \ldots f_r(\theta)^{e_r} \rangle \\ &\subseteq \langle p \rangle + \langle f(\theta) \rangle \\ &= \langle p \rangle. \end{aligned}$$

Thus $\langle p \rangle \mid \mathfrak{p}_1^{e_1} \ldots \mathfrak{p}_r^{e_r}$, and the only prime factors of $\langle p \rangle$ are $\mathfrak{p}_1, \ldots, \mathfrak{p}_r$, showing

$$\langle p \rangle = \mathfrak{p}_1^{k_1} \ldots \mathfrak{p}_r^{k_r} \qquad (1)$$

where

$$0 < k_i \leqslant e_i \qquad (1 \leqslant i \leqslant r). \qquad (2)$$

The norm of $\mathfrak{p}_i$ is, by definition, $|\mathfrak{O}/\mathfrak{p}_i|$, and by using the isomorphisms

$$\mathfrak{O}/\mathfrak{p}_i = \mathbf{Z}[\theta]/\mathfrak{p}_i \cong \mathbf{Z}_p[\theta_i]$$

we find

$$\mathrm{N}(\mathfrak{p}_i) = |\mathbf{Z}_p[\theta_i]| = p^{d_i}$$

where $d_i = \partial \bar{f}_i$, or equivalently, $d_i = \partial f_i$. Also

$$\mathrm{N}(\langle p \rangle) = |\mathbf{Z}[\theta]/\langle p \rangle| = p^n,$$

so, taking norms in (1), we find

$$p^n = \mathrm{N}(\langle p \rangle) = \mathrm{N}(\mathfrak{p}_1)^{k_1} \ldots \mathrm{N}(\mathfrak{p}_r)^{k_r} = p^{d_1 k_1 + \ldots d_r k_r},$$

which implies

$$d_1 k_1 + \ldots + d_r k_r = n = d_1 e_1 + \ldots + d_r e_r. \qquad (3)$$

From (2) we deduce $k_i = e_i$ $(1 \leqslant i \leqslant r)$ and this completes the proof. □

This result is not always applicable, since $\mathfrak{O}$ need not be of the form $\mathbf{Z}[\theta]$ in general. See Section 2.6, Example 2. But for quadratic or cyclotomic fields we have already shown that $\mathfrak{O} = \mathbf{Z}[\theta]$, so the theorem applies in these cases - and in many others. It also has the advantage of computability. Since there are only a finite number of polynomials over $\mathbf{Z}_p$ of given degree, the factorization of $\bar{f}$ can be performed in a finite number of steps. A little native wit helps, but, if the worst comes to the worst, there are only a finite number of polynomials of lower degree than $\bar{f}$ to try as factors.

For example, in $\mathbf{Q}(\sqrt{-1})$ we have $\mathfrak{O} = \mathbf{Z}[\theta]$ where θ has minimum polynomial

$$t^2 + 1.$$

To find the factorization of $\langle 2 \rangle$ we look at this polynomial mod 2, where we have

$$t^2 + 1 = (t+1)^2.$$

Hence $\langle 2 \rangle = \mathfrak{p}^2$ where

$$\begin{aligned}\mathfrak{p} &= \langle 2 \rangle + \langle \sqrt{-1} + 1 \rangle \\ &= \langle 1 + \sqrt{-1} \rangle\end{aligned}$$

(because $2 = (1 + \sqrt{-1})(1 - \sqrt{-1})$), and we recover the example noted at the beginning of this section.

More generally, consider the factorization in $\mathbf{Z}[\sqrt{-1}]$ of a prime $p \in \mathbf{Z}$. There are three cases to consider:

(i) $t^2 + 1$ irreducible mod p,

(ii) $t^2 + 1 = (t - \lambda)(t + \lambda) \bmod p$, where $\lambda^2 \equiv -1 \bmod p$) and $\lambda \neq -\lambda$ (i.e. $p \neq 2$),

(iii) $t^2 + 1 = (t+1)^2 \pmod 2$ when $p = 2$.

In case (i) $\langle p \rangle$ is prime; in case (ii) $\langle p \rangle = \mathfrak{p}_1 \mathfrak{p}_2$ for distinct prime ideals $\mathfrak{p}_1, \mathfrak{p}_2$; in case (iii) $\langle p \rangle = \mathfrak{p}_1^2$ for a prime ideal $\mathfrak{p}_1$.

The distinction between cases (i) and (ii) is whether or not -1 is congruent to a square (mod p). In the appendix on quadratic residues it may be seen that case (i) applies if p is of the form $4k - 1$ ($k \in \mathbf{Z}$), case (ii) if p is of the form $4k + 1$ ($k \in \mathbf{Z}$).

The results in this section are, in fact, the tip of the iceberg of a large and significant portion of algebraic number theory. Given a prime ideal $\mathfrak{p}$ in the ring $\mathfrak{O}_K$ of integers in a number field K, we may consider the extension ideal $\mathfrak{O}_L\mathfrak{p}$ in the ring of integers $\mathfrak{O}_L$ of an extension algebraic number field. We find

$$\mathfrak{O}_L \mathfrak{p} = \mathfrak{q}_1^{e1} \cdots \mathfrak{q}_r^{e_r}$$

where $\mathfrak{q}_1, \ldots, \mathfrak{q}_r$ are distinct prime ideals in $\mathfrak{O}_L$.

The study of the manner in which a general prime ideal factorizes when extended in this way will not detain us in this text, since the computational method we have given will prove more than satisfactory in our quest for a partial proof of Fermat's Last Theorem.

10.2 Minkowski's constants

The proof of Theorem 9.5 leaves room for improvement, in that it is based on Lemma 9.2, which is far stronger than we really need. What we want is a point α such that

$$|\sigma_1(\alpha)| \ldots |\sigma_s(\alpha)|\, |\sigma_{s+1}(\alpha)|^2 \ldots |\sigma_{s+t}(\alpha)|^2 < c_1 \ldots c_{s+t}; \tag{4}$$

but what we actually find is a point α satisfying the considerably greater restriction

$$\begin{gathered} |\sigma_1(\alpha)| < c_1, \ldots, |\sigma_s(\alpha)| < c_s, \\ |\sigma_{s+1}(\alpha)|^2 < c_{s+1}, \ldots, |\sigma_{s+t}(\alpha)|^2 < c_{s+t}. \end{gathered} \tag{5}$$

Certainly (5) implies (4), but not the reverse.

Our reason for using (5) is that we wish to employ Minkowski's theorem. For (5) the relevant set of points in $\mathbf{L}^{st}$ is convex and symmetric, so the theorem applies; but for (4) the relevant set, although symmetric, is not convex. This means we cannot use (5) directly. The gap between (4) and (5) is so great, however, that one might hope to find another set of inequalities, corresponding to a convex subset of $\mathbf{L}^{st}$, and implying (4): this would lead to improved estimates in Theorem 9.5 (and Corollary 9.6).

This can be done if we use the well-known inequality between the arithmetic and geometric means

$$(a_1 \ldots a_n)^{1/n} \leqslant \frac{1}{n}(a_1 + \ldots + a_n). \tag{6}$$

The result we obtain is:

Theorem 10.2. *If* $\mathfrak{a} \neq 0$ *is an ideal of* $\mathfrak{O}$ *then* $\mathfrak{a}$ *contains an element* α *with*

$$|N(\alpha)| \leqslant \left(\frac{4}{\pi}\right)^t \cdot \frac{n!}{n^n} \sqrt{|\Delta|} \, N(\mathfrak{a}),$$

where n is the degree of K and Δ *is the discriminant.*

Proof. Let X_c be the set of all $x \in \mathbf{L}^{st}$ such that

$$|x_1| + \ldots + |x_s| + 2\sqrt{(y_1^2 + z_1^2)} + \ldots + 2\sqrt{(y_t^2 + z_t^2)} < c,$$

where c is a positive real number. Then X_c is convex and centrally symmetric, and it is a routine though non-trivial exercise to compute

$$v(X_c) = 2^s \left(\frac{\pi}{2}\right)^t \cdot \frac{1}{n!} c^n$$

(using induction and a change to polar co-ordinates). For details of this computation see Lang [5] p. 116.

By Minkowski's theorem X_c contains a point $\alpha \neq 0$ of $\sigma(\mathfrak{a})$ provided

$$v(X_c) > 2^{s+2t} v(T),$$

where T is a fundamental domain for $\sigma(\mathfrak{a})$. By Theorem 9.4 we have

$$v(T) = 2^{-t} N(\mathfrak{a}) \sqrt{|\Delta|}$$

so the condition on X_c becomes

$$2^s \left(\frac{\pi}{2}\right)^t \cdot \frac{1}{n!} c^n > 2^{s+2t} 2^{-t} N(\mathfrak{a}) \sqrt{|\Delta|},$$

which is

$$c^n > \left(\frac{4}{\pi}\right)^t n! N(\mathfrak{a}) \sqrt{|\Delta|}.$$

For such an α we have

$$|N(\alpha)| = |\sigma_1(\alpha) \ldots \sigma_s(\alpha) \sigma_{s+1}(\alpha)^2 \ldots \sigma_{s+t}(\alpha)^2|$$

$$\leqslant \left(\frac{c}{n}\right)^n$$

by the inequality between arithmetic and geometric mean.

Using ϵ's as in Theorem 9.5 we may assume that α can be found for

$$c^n = \left(\frac{4}{\pi}\right)^t n!\mathrm{N}(\mathfrak{a})\sqrt{|\Delta|}$$

and then

$$|\mathrm{N}(\alpha)| \leqslant \left(\frac{4}{\pi}\right)^t \cdot \frac{n!}{n^n} \mathrm{N}(\mathfrak{a})\sqrt{|\Delta|}. \qquad \square$$

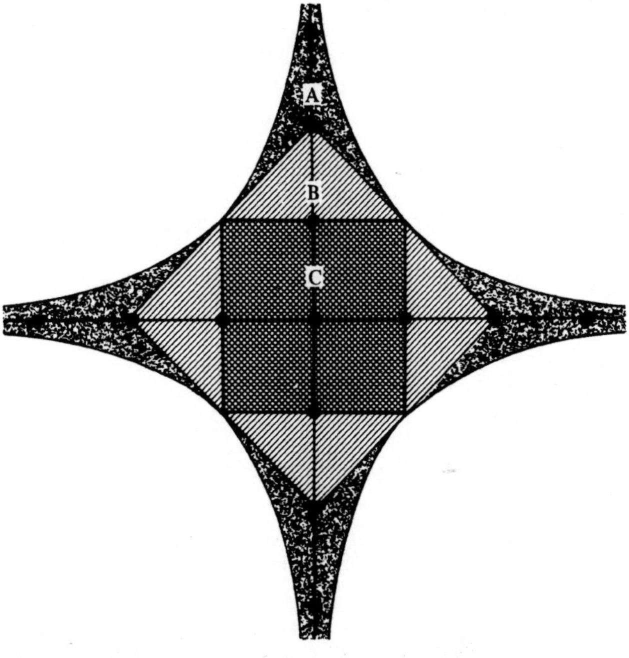

Fig. 9. Geometry suggests the choice of X_c in the proof of Theorem 10.2, here illustrated for $n = s = 2, t = 0$. Region B is convex, lies within region A, and is larger than the more obvious region C: in consequence the use of B, in conjunction with Minkowski's theorem, yields a better bound.

The geometric considerations involved in the choice of X_c in this proof are illustrated in Fig. 9 for the case where $n = 2$, $s = 2$, $t = 0$. The three regions

A: $|xy| \leqslant 1$

B: $\dfrac{|x| + |y|}{2} \leqslant 1$

C: $|x| \leqslant 1, \quad |y| \leqslant 1$

correspond respectively to the inequality (4), the region chosen in the proof of Theorem 10.2, and the inequality (5). Note that A is not convex, although B, C are; that $C \subseteq B \subseteq A$; and that B is much larger than C (which is why it leads to a better estimate).

Corollary 10.3. *Every class of fractional ideals contains an ideal* $\mathfrak{a}$ *with*

$$N(\mathfrak{a}) \leqslant \left(\frac{4}{\pi}\right)^t \cdot \frac{n!}{n^n} \sqrt{|\Delta|}.$$

Proof. As for Corollary 9.6. □

This result suggests the introduction of *Minkowski constants*

$$M_{st} = \left(\frac{4}{\pi}\right)^t \frac{(s+2t)!}{(s+2t)^{s+2t}}.$$

For future use, we give a short table of their values, taken from Lang [6].

n	s	t	M_{st}
2	0	1	0.637
2	2	0	0.500
3	1	1	0.283
3	3	0	0.223
4	0	2	0.152
4	2	1	0.120
4	4	0	0.094
5	1	2	0.063
5	3	1	0.049
5	5	0	0.039

(The numbers in the last column have all been rounded *upwards* in the third decimal place, to avoid under-estimates.)

We can now give a criterion for a number field to have class-number 1, for which the calculations required are often practicable.

Theorem 10.4. *Let $\mathfrak{O}$ be the ring of integers of a number field K of degree $n = s + 2t$. Suppose that for every prime $p \in \mathbf{Z}$ with*

$$p \leqslant M_{st}\sqrt{|\Delta|}$$

(Δ being the discriminant of K), every prime ideal dividing $\langle p \rangle$ is principal. Then $\mathfrak{O}$ has class-number $h = 1$.

Proof. Every class of fractional ideals contains an ideal $\mathfrak{a}$ with $N(\mathfrak{a}) \leqslant M_{st}\sqrt{|\Delta|}$. Now

$$N(\mathfrak{a}) = p_1 \ldots p_k$$

where $p_1, \ldots, p_k \in \mathbf{Z}$ and $p_i \leqslant M_{st}\sqrt{|\Delta|}$; and $\mathfrak{a} \mid N(\mathfrak{a})$, so $\mathfrak{a}$ is a product of prime ideals, each dividing some p_i. By hypothesis these prime ideals are principal, so $\mathfrak{a}$ is principal. Therefore every class of fractional ideals is equal to $[\mathfrak{O}]$, and $h = 1$. □

Specific numerical applications of this theorem, and related methods, follow in the next section.

10.3 Some class-number calculations

Theorem 10.4 combines with Theorem 10.1 to provide a useful computational technique for fields of small degree and with small discriminant. The following examples show what is meant by 'small' in these circumstances.

1. $\mathbf{Q}(\sqrt{-19})$: The ring of integers is $\mathbf{Z}[\theta]$ where θ is a zero of

$$f(t) = t^2 - t + 5,$$

and the discriminant is 19. Then $M_{st}\sqrt{|\Delta|} \leqslant 0.637\sqrt{19}$ so Theorem 10.4 applies if we know the factors of primes $\leqslant 3$. Now we use Theorem 10.1: modulo 2, $f(t)$ is irreducible, so $\langle 2 \rangle$ is prime in $\mathfrak{O}$ (and hence every prime ideal dividing $\langle 2 \rangle$ is equal to $\langle 2 \rangle$ so is principal); modulo 3, $f(t)$ is also irreducible, so $\langle 3 \rangle$ is prime and the same argument applies.

2. $\mathbf{Q}(\sqrt{-43})$: This is similar, but now

$$f(t) = t^2 - t + 11$$

and $M_{st}\sqrt{|\Delta|} \leqslant 0.637\sqrt{43}$ which involves looking at primes $\leqslant 4$. But $f(t)$ is irreducible modulo 2 or 3.

3. $\mathbf{Q}(\sqrt{-67})$: For this,

$$f(t) = t^2 - t + 17$$

and $M_{st}\sqrt{|\Delta|} \leqslant 0.637\sqrt{67}$ which involves looking at primes $\leqslant 5$. But $f(t)$ is irreducible modulo 2, 3 or 5.

4. $\mathbf{Q}(\sqrt{-163})$: Now

$$f(t) = t^2 - t + 41$$

and $M_{st}\sqrt{|\Delta|} \leqslant 0.637\sqrt{163}$ which involves looking at primes $\leqslant 8$. But $f(t)$ is irreducible modulo 2, 3, 5, or 7.

Combining these results with Theorem 4.17 (or using the above methods for the other values of Δ) we have:

Theorem 10.5. *The class-number of* $\mathbf{Q}(\sqrt{d})$ *is equal to* 1 *for* $d = -1, -2, -3, -7, -11, -19, -43, -67, -163$. □

As we have remarked in Section 4.3, these are in fact the only values of $d < 0$ for which $\mathbf{Q}(\sqrt{d})$ has unique factorization, or equivalently class-number 1.

Comparing the Theorem 4.18 we obtain the interesting:

Corollary 10.6. *There exist rings which have unique factorization but are not Euclidean; for example the rings of integers of* $\mathbf{Q}(\sqrt{d})$ *for* $d = -19, -43, -67, -163$. □

We can also deal with a few cyclotomic fields by the same method. If $K = \mathbf{Q}(\zeta)$ where $\zeta^p = 1$, p prime, then the degree of K is $p - 1$, and the ring of integers is $\mathbf{Z}[\zeta]$. For $p = 3$, $K = \mathbf{Q}(\sqrt{-3})$ and we already know $h = 1$ for this.

5. $\mathbf{Q}(\zeta)$ where $\zeta^5 = 1$: Here $n = 4, s = 0, t = 2$; and $\Delta = 125$

by Theorem 3.6. Hence $M_{st}\sqrt{|\Delta|} \leqslant 0.152\sqrt{125}$ so we must look at primes $\leqslant 1$. Since there are no such primes, Theorem 10.4 applies at once to give $h = 1$.

6. $\mathbf{Q}(\zeta)$ where $\zeta^7 = 1$: Here $n = 6, s = 0, t = 3$, and $\Delta = -7^5$. We have to look at primes $\leqslant 3$. The ring of integers is $\mathbf{Z}[\zeta]$ where ζ is a zero of

$$f(t) = t^6 + t^5 + t^4 + t^3 + t^2 + t + 1.$$

Modulo 2, this factorizes as

$$(t^3 + t^2 + 1)(t^3 + t + 1)$$

so $\langle 2 \rangle = \mathfrak{p}_1 \mathfrak{p}_2$ where $\mathfrak{p}_1, \mathfrak{p}_2$ are distinct prime ideals, by Theorem 10.1. In fact

$$(\zeta^3 + \zeta^2 + 1)(\zeta^3 + \zeta + 1)\zeta^4 = 2,$$

so we have

$$\langle 2 \rangle = \langle \zeta^3 + \zeta^2 + 1 \rangle \langle \zeta^3 + \zeta + 1 \rangle$$

and $\mathfrak{p}_1, \mathfrak{p}_2$ are principal.

Modulo 3, $f(t)$ is irreducible (by trying all possible divisors, or more enlightened methods), so $\langle 3 \rangle$ is prime.

Hence by Corollary 10.4, $h = 1$.

Similar methods often allow us to compute h, even when it is not 1.

7. $\mathbf{Q}(\sqrt{10})$: The discriminant $d = 40, n = 2, s = 2, t = 0$. Every class of ideals contains one with norm

$$\leqslant M_{2,0}\sqrt{|\Delta|} \leqslant 0.5\sqrt{40}$$

so we must factorize the primes $\leqslant 3$. Now $\mathfrak{O} = \mathbf{Z}[\theta]$ where θ is a zero of

$$f(t) = t^2 - 10.$$

$f(t) \equiv (t + 1)(t - 1) \pmod 3$, so $\langle 3 \rangle = \mathfrak{g}_1 \mathfrak{g}_2$ where $\mathfrak{g}_1 = \langle 3, 1 + \sqrt{10} \rangle$, $\mathfrak{g}_2 = \langle 3, 1 - \sqrt{10} \rangle$. Modulo 2 we have $f(t) = t.t$ so that $\langle 2 \rangle = \mathfrak{p}^2$ for a prime ideal $\mathfrak{p}$. If $\mathfrak{p}$ is principal, say $\mathfrak{p} = \langle a + b\sqrt{10} \rangle$, then the equation

$$\mathrm{N}(p)^2 = \mathrm{N}(\langle 2 \rangle) = 4$$

implies that $N(\mathfrak{p}) = 2$, Hence

$$a^2 - 10b^2 = \pm 2.$$

The latter, considered modulo 10, is impossible; hence $\mathfrak{p}$ is not principal.

We have $\mathfrak{p}\mathfrak{g}_1 = \langle -2 + \sqrt{10} \rangle$ and $[\mathfrak{g}_1] = [\mathfrak{p}]^{-1}$.

This means that every class of fractional ideals either contains a principal ideal or $\mathfrak{p}$, hence equals $[\mathfrak{O}]$ or $[\mathfrak{p}]$. Since $\mathfrak{p}$ is not principal, these two classes are distinct, so $h = 2$. The class-group is cyclic of order 2, and as verification we have

$$[\mathfrak{p}]^2 = [\mathfrak{p}^2] = [\langle 2 \rangle] = [\mathfrak{O}].$$

As we said in Section 4.4, all the imaginary quadratic fields $\mathbf{Q}(\sqrt{d})$ with unique factorization are now known, verifying a conjecture of Gauss. But Gauss also stated a more general conjecture, the *Class Number Problem.* This states that for any given class number h, the set of $d < 0$ for which $\mathbf{Q}(\sqrt{d}) = h$ is finite. It was proved in 1983 by Goldfeld, Gross, and Zagier, and is described in a masterly survey by Goldfeld [25].

10.4 Tables

To give an idea of how irregularly the class-number of $\mathbf{Q}(\sqrt{d})$ depends upon d, we give a short table showing, for squarefree d with $0 < d < 100$, the class-numbers h of $\mathbf{Q}(\sqrt{d})$ and h' of $\mathbf{Q}(\sqrt{-d})$.

d	h	h'	d	h	h'	d	h	h'
1	–	1	34	2	4	69	1	8
2	1	1	35	2	2	70	2	4
3	1	1	37	1	2	71	1	7
5	1	2	38	1	6	73	1	4
6	1	2	39	2	4	74	2	10
7	1	1	41	1	8	77	1	8
10	2	2	42	2	4	78	2	4
11	1	1	43	1	1	79	3	5
13	1	2	46	1	4	82	4	4
14	1	4	47	1	5	83	1	3
15	2	2	51	2	2	85	2	4
17	1	4	53	1	6	86	1	10
19	1	1	55	2	4	87	2	6
21	1	4	57	1	4	89	1	12
22	1	2	58	2	2	91	2	2
23	1	3	59	1	3	93	1	4
26	2	6	61	1	6	94	1	8
29	1	6	62	1	8	95	2	8
30	2	4	65	2	8	97	1	4
31	1	3	66	2	8			
33	1	4	67	1	1			

Methods more suited to such computations than ours above exist, especially analytic methods which are beyond our present scope. See Borevič and Šafarevič [1] p. 342 ff.

Exercises

10.1 Let $K = \mathbb{Q}(\sqrt{3})$. Use Theorem 10.1 to factorize the following principal ideals in the ring $\mathfrak{O}$ of integers of K:

$$\langle 2\rangle, \langle 3\rangle, \langle 5\rangle, \langle 10\rangle, \langle 30\rangle.$$

10.2 Factorize the following principal ideals in the ring of integers of $\mathbb{Q}(\sqrt{5})$:

$$\langle 2\rangle, \langle 3\rangle, \langle 5\rangle, \langle 12\rangle, \langle 25\rangle.$$

10.3 Factorize the following ideals in $\mathbf{Z}[\zeta]$ where $\zeta = e^{2\pi i/5}$:

$$\langle 2\rangle, \langle 5\rangle, \langle 20\rangle, \langle 50\rangle.$$

10.4 Compute the volume integral quoted in the proof of Theorem 10.2.

10.5 If K is a number field of degree n, prove that

$$|\Delta| \geqslant \left(\frac{\pi}{4}\right)^n \left(\frac{n^n}{n!}\right)^2,$$

where Δ is the discriminant.

10.6 Prove that there can exist only finitely many number fields with any given discriminant.

10.7 Using the methods of this chapter, compute the class-numbers of fields $\mathbf{Q}(\sqrt{d})$ for $-20 \leqslant d \leqslant 20$.

CHAPTER ELEVEN

Fermat's Last Theorem

We now have enough machinery at our disposal to tackle Fermat's Last Theorem in a special case, namely when the exponent n in the equation $x^n + y^n = z^n$ is a so-called 'regular' prime, and when n does not divide any of x, y, or z. We begin with a short historical survey to set the problem in perspective. Following this we show how elementary methods dispose of the case $n = 4$, and reduce the problem to odd prime values of n. Then, following Kummer and Dedekind, we apply the methods of ideal theory to the aforementioned special case. We cannot in this book deal with the situation where one of x, y, or z is divisible by n; and even less can we discuss irregular primes; but the power of the method is clear even in the restricted situation accessible to us. In a final discursive section without proofs we discuss the regularity property and some related matters.

11.1 Some history

The origins of Fermat's Last Theorem have been discussed in the introduction to this book. Fermat himself disposed of the cases $n = 3, 4$; as did Euler independently later. The case $n = 5$ was proved by Dirichlet in 1828 and Legendre in 1830; Gauss left a posthumous sketch proof which he noted would not extend to cover $n = 7$. In 1832 Dirichlet proved the conjecture for $n = 14$. Lamé offered a proof for $n = 7$ in 1839; his errors were pointed out by Lebesgue in 1840 who gave a corrected and simplified proof. In 1847 Lamé claimed a proof for all n, but made the standard error of assuming unique factorization, as was pointed out immediately by Liouville and Kummer. Cauchy made a more interesting error: he claimed to have proved the uniqueness, but finally admitted at the end of a series of papers that his proof failed for $n = 23$ (see [20] p. 308).

In 1850 came Kummer's sensational proof for 'regular' primes, including all primes less than 100 except for 37, 59, 67. Kummer asserted that there were infinitely many regular primes but this has never been proved (although it is known that there are infinitely many *irregular* primes). The same year he disposed of these three cases – but made errors which went undetected until 1920 when Vandiver noticed them. A proof for $n = 37$ was given by Mirimanoff in 1893, and in 1905 he pushed on as far as $n \leqslant 257$. Vandiver laid down methods which made a computational approach possible, so that now we know the conjecture to be true for all $n \leqslant 125\,000$, see [36, 42].

Meanwhile a different approach was developed by Legendre in 1823, to handle a special case of the conjecture (the so-called 'first case') in which x, y, z are prime to n. He proved that n cannot be of the form $2p + 1, 4p + 1, 8p + 1, 10p + 1, 14p + 1$, or $16p + 1$, where both n and p are prime; except perhaps for $n = 31, 43$. Dickson in 1908 proved a similar result for $n = kp + 1$ where $k = 20, 22, 26, 28, 32, 40, 56, 64$ (with a few small exceptions), from which eventually he proved that n cannot be a prime $\leqslant 7000$.

More decisive criteria came from the work of Wieferich, who in 1909 showed that in the first case of the conjecture, if a solution exists then

$$2^n \equiv 2 \pmod{n^2}$$

(n prime as usual). Other writers subsequently proved similar results, of the form

$$q^n \equiv q \qquad (\text{mod } n^2)$$

for $q = 3, 5, 7, 11, 13, 17, 19, 23, 29, 31, 37, 41, 43$ (Mirimanoff, Frobenius, Vandiver, Pollackzek, Morishima, Rosser). Using these results the Lehmers proved the first case of the conjecture true for all prime $n < 253747889$. This has been improved to 3×10^9 by Brillhart, Tonascia and Weinberger [17].

It will be evident that the numbers involved have become so large that there is no chance of exhibiting a counterexample *directly*. In fact, if a counterexample exists, then x must have at least 18×10^5 digits. See Ribenboim [10] p. 24. This highly readable book contains a wealth of information on Fermat's Last Theorem. This is not to say that a negative solution to the conjecture is impossible.

The most dramatic development in recent times is the proof, by Gerd Faltings in 1983, that for any $n \geqslant 3$ there are at most a finite number of solutions (with x, y, z having no common factor) to the Fermat equation. Faltings derived this result as a corollary to a far more general result, known as the Mordell Conjecture. Associated with any homogeneous equation $f(x, y, z) = 0$ is a number called its *genus*. In 1922 Louis Mordell conjectured that any equation with genus $\geqslant 2$ has only finitely many solutions (with x, y, z pairwise coprime). The Fermat equation has genus $\frac{1}{2}(n-1)(n-2)$ which is greater than 1 when $n > 3$, so Faltings' Theorem applies. For an accessible discussion see Bloch [18].

One may ask whether, in view of all the effort expended by so many mathematicians, without complete success, Fermat's claim to a proof is credible. This is a vexed question and, in the absence of any hard evidence, pure speculation. However, it is perhaps worth a paragraph. Fermat's judgement of the correctness of his proofs seems to have been very good, especially by the standards of the time; and his intuitive understanding of numbers was exceptional. On the other hand, even he would most probably have failed to appreciate such questions as uniqueness of factorization of algebraic numbers; and 'proofs' based on such methods were in use at the time of Fermat, or soon after. It is thus quite possible that his proof rested on some assumption whose contentious nature was not evident at the time. At least one of Fermat's conjectures has proved to be incorrect, namely that the numbers of the form $F_n = 2^{2^n} + 1$ for non-negative n are all prime. The first few, $F_0 = 3, F_1 = 5, F_2 = 17, F_3 = 257$, $F_4 = 65537$ were known to be prime by Fermat, but F_5 proves to have 641 as a factor.(To Fermat's credit, he did not claim a proof of this conjecture.) It is thus within the bounds of possibility that Fermat's Last Theorem is also false.

In the history of human thought there is a strong tradition whereby the 'ancients' are credited with superhuman powers which have been lost – usually on the flimsiest of evidence – and one cannot but form the impression that this is what many *wish* to believe. It is conceivable, but unlikely, that Fermat found a very clever trick to prove his conjecture. If so, the work of 300 years has not revealed what it might be. But the experience of these 300 years is that the whole question is one of enormous depth, and that Fermat's Last Theorem is closely linked to tremendously difficult problems about algebraic numbers. Whether the theorem eventually proves to be true or false, we shall never be sure of the nature of Fermat's proof. It seems that it is likely that, as Struik [40] says, 'even the great Fermat slept sometimes'.

11.2 Elementary considerations

We consider what can be said about the *Fermat equation*

$$x^n + y^n = z^n \tag{1}$$

from an elementary point of view. If a solution to (1) exists then there must exist one in which x, y, z are coprime in pairs. For if a prime q divides x and y, then $x = qx', y = qy'$,

$$q^n(x'^n + y'^n) = z^n$$

so that q also divides z, say $z = qz'$, and then $x'^n + y'^n = z'^n$. Similarly if q divides x and z, or y and z. In this way we can remove all common factors from x, y, z.

Next note that if (1) is impossible for an exponent n then it is impossible for all multiples of n. For if $x^{mn} + y^{mn} = z^{mn}$ then $(x^m)^n + (y^m)^n = (z^m)^n$. Now any integer $\geqslant 3$ is divisible either by 4 or by an odd prime. Hence to prove (or disprove) the conjecture *it is sufficient to consider the cases $n = 4$ and n an odd prime.*

We start with Fermat's proof for $n = 4$. It is based on the (well known) general solution of the Pythagorean equation $x^2 + y^2 = z^2$, given by:

Lemma 11.1. *The solutions of $x^2 + y^2 = z^2$ with pairwise coprime integers x, y, z are given parametrically by*

$$\begin{aligned} \pm x &= r^2 - s^2 \\ \pm y &= 2rs \\ \pm z &= r^2 + s^2 \end{aligned}$$

(or with x, y interchanged) where r, s are coprime and exactly one is odd.

Proof. We shall give the classical proof. It is sufficient to consider x, y, z to be positive. They cannot all be odd, for this gives the contradiction 'odd + odd = odd'. Since they are pairwise coprime, this means precisely one is even. It cannot be z, for then $z = 2k$, $x = 2a + 1$, $y = 2b + 1$ where k, a, b are rational integers, and

$$(2a + 1)^2 + (2b + 1)^2 = 4k^2.$$

This cannot occur since the left-hand side is clearly not divisible by 4 whilst the right-hand side is. So one of x, y is even. We can suppose that this is x. Then

$$x^2 = z^2 - y^2 = (z + y)(z - y).$$

Because $x, z + y, z - y$ are all even and positive, we can write $x = 2u$, $z + y = 2v$, $z - y = 2w$, whence

$$(2u)^2 = 2v.2w,$$

or

$$u^2 = vw. \tag{2}$$

Now v, w are coprime, for a common factor of v, w would divide their sum $v + w = z$ and their difference $v - w = y$, which have no proper common factors, Factorizing u, v, w into prime factors, we see that (2) implies v, w are both squares, say $v = r^2$, $w = s^2$. Moreover r, s are coprime because v, w are.

Thus

$$z = v + w = r^2 + s^2,$$

$$y = v - w = r^2 - s^2.$$

Because y, z are both odd, precisely one of r, s is odd. Finally

$$x^2 = z^2 - y^2 = (r^2 + s^2)^2 - (r^2 - s^2)^2 = 4r^2s^2,$$

so

$$x = 2rs. \qquad \square$$

Now we can prove a theorem even stronger than the impossibility of (1) for $n = 4$, namely:

Theorem 11.2. *The equation $x^4 + y^4 = z^2$ has no integer solutions with $x, y, z \neq 0$.*

Proof. First note that this *is* stronger, since if $x^4 + y^4 = z^4$ then x, y, z^2 satisfy the above equation.

Suppose a solution of

$$x^4 + y^4 = z^2 \tag{3}$$

exists. We may assume x, y, z are positive. Among such solutions there exists one for which z is smallest: assume we have this one in (3). Then x, y, z are coprime (or we can cancel a common factor and make z smaller) and so by Lemma 11.1 we have

$$x^2 = r^2 - s^2, \qquad y^2 = 2rs, \qquad z = r^2 + s^2,$$

where x, z are odd and y is even. The first of these implies

$$x^2 + s^2 = r^2$$

with x, s coprime. Hence by 11.1 again, since x is odd

$$x = a^2 - b^2, \qquad s = 2ab, \qquad r = a^2 + b^2.$$

But now we substitute back to get

$$y^2 = 2rs = 2.2ab(a^2 + b^2)$$

so y is even, say $y = 2k$, and

$$k^2 = ab(a^2 + b^2).$$

Since a, b and $a^2 + b^2$ are pairwise coprime we have

$$a = c^2, \qquad b = d^2, \qquad a^2 + b^2 = e^2,$$

so that

$$c^4 + d^4 = e^2.$$

This is an equation of type (3), but $e \leqslant a^2 + b^2 = r < z$, so we have contradicted the minimality of z. □

11.3 Kummer's lemma

This section begins our build-up to the solution of a special case of Fermat's Last Theorem, with a detailed study of the field $K = \mathbf{Q}(\zeta)$ where $\zeta = e^{2\pi i/p}$ for an odd prime p. As in Chapter 3 we write

$$\lambda = 1 - \zeta.$$

Further we define

$$\mathfrak{l} = \langle \lambda \rangle,$$

the ideal generated by λ in the ring of integers $\mathbf{Z}[\zeta]$ of K. We start with some properties of $\mathfrak{l}$.

Lemma 11.3. (a) $\mathfrak{l}^{p-1} = \langle p \rangle$, (b) $N(\mathfrak{l}) = p$.

Proof. First note that for $j = 1, \ldots, p-1$ the numbers $1 - \zeta$ and $1 - \zeta^j$ are associates in $\mathbf{Z}[\zeta]$. Clearly $1 - \zeta \mid 1 - \zeta^j$. But if we choose t such that $jt \equiv 1 \pmod p$ then $1 - \zeta = 1 - \zeta^{jt}$ so that $1 - \zeta^j \mid 1 - \zeta$. Hence they are associates.

Now Equation (9) of Chapter 3 leads to

$$\langle p \rangle = \prod_{j=1}^{p-1} \langle 1 - \zeta^j \rangle$$

but the above remarks show that $\langle 1 - \zeta^j \rangle = \langle 1 - \zeta \rangle = \mathfrak{l}$, so that

$$\langle p \rangle = \mathfrak{l}^{p-1}$$

and (a) is proved. Part (b) is immediate on taking norms. □

A useful consequence of (b) should be noted. It implies that $|\mathbf{Z}[\zeta]/\mathfrak{l}| = p$, from which it follows (on looking at the natural homomorphism $\mathbf{Z}[\zeta] \to \mathbf{Z}[\zeta]/\mathfrak{l}$) that every element of $\mathbf{Z}[\zeta]$ is congruent modulo $\mathfrak{l}$ to one of $0, 1, 2, \ldots, p-1$.

The main aim of the rest of this section is to give a useful, though incomplete, description of the units of $\mathbf{Z}[\zeta]$. We start by finding which roots of unity occur, showing that there are no 'accidental' occurrences:

Lemma 11.4. *The only roots of unity in K are $\pm \zeta^s$ for integers s.*

Proof. First we show $i \notin K$. If $i \in K$ then since $2 = i(1-i)^2$ we have

$$\langle 2 \rangle = \langle 1 - i \rangle^2.$$

Hence when $\langle 2 \rangle$ is resolved into prime factors in $\mathbf{Z}[\zeta]$ it has repeated factors. By Theorem 10.1 this implies that the polynomial

$$f(t) = \frac{t^p - 1}{t - 1}$$

has a repeated irreducible factor modulo 2, hence that $t^p - 1$ has a repeated irreducible factor modulo 2. Then the remark following Theorem 1.2 tells us that $t^p - 1$ and $\mathrm{D}(t^p - 1) = pt^{p-1}$ are not coprime. However, p is odd, so these polynomials modulo 2 take the form $t^p + 1$, t^{p-1} which are obviously coprime. This is a contradiction.

In exactly the same way we can show that for an odd

prime $q \neq p$,

$$e^{2\pi i/q} \notin K.$$

We just use

$$\langle q \rangle = \langle 1 - e^{2\pi i/q} \rangle^{q-1}.$$

Next we remark that

$$e^{2\pi i/p^2} \notin K.$$

For $e^{2\pi i/p^2}$ satisfies $t^{p^2} - 1 = 0$, but not $t^p - 1 = 0$ and so is a root of

$$f(t) = (t^{p^2} - 1)/(t^p - 1) = \sum_{r=0}^{p-1} t^{rp} = 0.$$

By applying Eisenstein's criterion to $f(t+1)$, a little arithmetic shows $f(t+1)$, and hence $f(t)$, is irreducible. Thus f is the minimum polynomial of $e^{2\pi i/p^2}$. Since $[K:\mathbb{Q}] = p - 1$ it follows from Theorems 1.7 and 1.8 that $e^{2\pi i/p^2} \notin K$.

Suppose now that $e^{2\pi i/m} \in K$ for an integer m. Then the above results show that

$$4 \nmid m, \qquad q \nmid m, \qquad p^2 \nmid m.$$

Hence $m \mid 2p$ which leads at once to the desired result. □

Lemma 11.5. *For each* $\alpha \in \mathbf{Z}[\zeta]$ *there exists* $a \in \mathbf{Z}$ *such that* $\alpha^p \equiv a \pmod{\mathfrak{l}^p}$.

Proof. We have already remarked on the existence of $b \in \mathbf{Z}$ such that $\alpha \equiv b \pmod{\mathfrak{l}}$. Now

$$\alpha^p - b^p = \prod_{j=0}^{p-1} (\alpha - \zeta^j b)$$

and since $\zeta \equiv 1 \pmod{\mathfrak{l}}$ each factor on the right is congruent to $\alpha - b \equiv 0 \pmod{\mathfrak{l}}$. Multiplying up, $\alpha^p - b^p \equiv 0 \pmod{\mathfrak{l}^p}$. □

Next comes a curious result about polynomials and roots of unity:

Lemma 11.6. *If $p(t) \in \mathbf{Z}[t]$ is a monic polynomial, all of whose zeros in $\mathbf{C}$ have absolute value 1, then every zero is a root of unity.*

Proof. Let $\alpha_1, \ldots, \alpha_k$ be the zeros of $p(t)$. For each integer $l > 0$ the polynomial

$$p_l(t) = (t - \alpha_1^l) \ldots (t - \alpha_k^l)$$

lies in $\mathbf{Z}[t]$ by the usual argument on symmetric polynomials. Now if

$$p_l(t) = t^k + a_{k-1}t^{k-1} + \ldots + a_0$$

then

$$|a_j| \leqslant \binom{k}{j} \qquad (j = 0, \ldots, k-1)$$

by estimating the size of elementary symmetric polynomials in the α_j and using $|\alpha_j| = 1$. But only finitely many distinct polynomials over $\mathbf{Z}$ can satisfy this system of inequalities, so for some $m \neq l$ we must have

$$p_l(t) = p_m(t).$$

Hence there exists a permutation π of $\{1, \ldots, k\}$ such that

$$\alpha_j^l = \alpha_{\pi(j)}^m$$

for $j = 1, \ldots, k$. Inductively we find that

$$\alpha_j^{l^r} = \alpha_{\pi^r(j)}^{m^r}$$

and so, since $\pi^{k!}(j) = j$, we have $\alpha_j^{l^{k!}} = \alpha_j^{m^{k!}}$ and hence

$$\alpha_j^{(l^{k!} - m^{k!})} = 1.$$

Since $l^{k!} \neq m^{k!}$ it follows that α_j is a root of unity. □

Now we may prove the main result of this section, known as *Kummer's lemma*:

Lemma 11.7. *Every unit of $\mathbf{Z}[\zeta]$ is of the form $r\zeta^g$ where r is real and g is an integer.*

Proof. Let ϵ be a unit in $\mathbf{Z}[\zeta]$. There exists a polynomial $e(t) \in \mathbf{Z}[t]$ such that $\epsilon = e(\zeta)$. For $s = 1, \ldots, p-1$ we have

$$\epsilon_s = e(\zeta^s)$$

conjugate to ϵ. Now $1 = \pm N(\epsilon) = \pm \epsilon_1 \ldots \epsilon_{p-1}$, so each ϵ_s is also a unit. Further, if bars denote complex conjugation, we have

$$\epsilon_{p-s} = e(\zeta^{p-s}) = e(\zeta^{-s}) = e(\overline{\zeta^s}) = \overline{e(\zeta^s)} = \overline{\epsilon_s}.$$

Therefore

$$\epsilon_s \epsilon_{p-s} = |\epsilon_s|^2 > 0.$$

Then

$$\pm 1 = N(\epsilon) = (\epsilon_1 \epsilon_{p-1})(\epsilon_2 \epsilon_{p-2}) \ldots > 0$$

so that $N(\epsilon) = 1$.

Now each $\epsilon_s/\epsilon_{p-s}$ is a unit, of absolute value 1, and by a symmetric polynomial argument

$$\prod_{s=1}^{p-1} (t - \epsilon_s/\epsilon_{p-s})$$

has coefficients in $\mathbf{Z}$. By Lemma 11.6 its zeros are roots of unity. An appeal to Lemma 11.4 yields the equation

$$\epsilon/\epsilon_{p-1} = \pm \zeta^u$$

for integer u. Since p is odd either u or $u + p$ is even, so we have

$$\epsilon/\epsilon_{p-1} = \pm \zeta^{2g} \qquad (4)$$

for $0 < g \in \mathbf{Z}$.

The crucial step now is to find out whether the sign in (4) is positive or negative. To do this we work out the left-hand side modulo $\mathfrak{l}$, as follows: we know that for some $v \in \mathbf{Z}$

$$\zeta^{-g}\epsilon \equiv v \pmod{\mathfrak{l}}$$

whence by taking complex conjugates

$$\zeta^g \epsilon_{p-1} \equiv v \pmod{\langle \bar{\lambda} \rangle}.$$

But $\bar{\lambda} = 1 - \zeta^{p-1}$ is an associate of λ, so in fact $\langle \bar{\lambda} \rangle = \mathfrak{l}$. Eliminating v leads to

$$\epsilon/\epsilon_{p-1} \equiv \zeta^{2g} \pmod{\mathfrak{l}}.$$

With a negative sign in (4) we are led to

$$\mathfrak{l} \mid 2\zeta^{2g}$$

and hence, taking norms,

$$N(\mathfrak{l}) \mid 2^{p-1}$$

which contradicts Lemma 11.3(b). So the sign in (4) is positive. Hence

$$\zeta^{-g}\epsilon = \zeta^{g}\epsilon_{p-1}.$$

The two sides of this equation are complex conjugates, so are in fact real; hence $\zeta^{-g}\epsilon = r \in \mathbf{R}$ which proves the lemma.

□

11.4 Kummer's Theorem

In order to state Kummer's special case of Fermat's Last Theorem, we need a technical definition. A prime p is said to be *regular* if it does not divide the class-number of $\mathbf{Q}(\zeta)$, where $\zeta = e^{2\pi i/p}$. By Section 10.3, $p = 3, 5, 7$ are regular. Further discussion of the regularity property is postponed until Section 11.5, for we are now in a position to state and prove:

Theorem 11.8. *If p is an odd regular prime then the equation*

$$x^p + y^p = z^p$$

has no solutions in integers x, y, z satisfying

$$p \nmid x, \qquad p \nmid y, \qquad p \nmid z.$$

Proof. We consider instead the equation

$$x^p + y^p + z^p = 0 \tag{5}$$

which exhibits greater symmetry. Since we can pass from this to the Fermat equation by changing z to $-z$, it suffices to work on (5). We assume, for a contradiction, that there exists a solution (x, y, z) of (5) in integers prime to p. We may as usual assume further that x, y, z are pairwise coprime. We may factorize (5) in $\mathbf{Q}(\zeta)$ to obtain

$$\prod_{j=0}^{p-1}(x + \zeta^j y) = -z^p$$

and pass to ideals:

$$\prod_{j=0}^{p-1}\langle x + \zeta^j y\rangle = \langle z\rangle^p. \tag{6}$$

First we establish that all factors on the left of this equation are pairwise coprime. For suppose $\mathfrak{p}$ is a prime ideal dividing $\langle x + \zeta^k y\rangle$ and $\langle x + \zeta^l y\rangle$ with $0 \leqslant k < l \leqslant p - 1$. Then $\mathfrak{p}$ contains

$$(x + \zeta^k y) - (x + \zeta^l y) = y\zeta^k(1 - \zeta^{l-k}).$$

Now $1 - \zeta^{l-k}$ is an associate of $1 - \zeta = \lambda$, and ζ^k is a unit, so $\mathfrak{p}$ contains $y\lambda$. Since $\mathfrak{p}$ is prime either $\mathfrak{p} \mid y$ or $\mathfrak{p} \mid \lambda$. In the first case $\mathfrak{p}$ also divides z from (6). Now y and z are coprime integers, so there exist $a, b \in \mathbf{Z}$ such that $az + by = 1$. But $y, z \in \mathfrak{p}$ so $1 \in \mathfrak{p}$, a contradiction. On the other hand, since $N(\mathfrak{l}) = p$ it follows from Theorem 5.11(a) that $\mathfrak{l}$ is prime; so if $\mathfrak{p} \mid \lambda$ then $\mathfrak{p} = \mathfrak{l}$. Then $\mathfrak{l} \mid z$ so

$$p = N(\mathfrak{l}) \mid N(z) = z^{p-1}$$

and $p \mid z$ contrary to hypothesis.

Uniqueness of prime factorization of ideals now implies that each factor on the left of (6) is a pth power of an ideal, since the right-hand side is a pth power and the factors are pairwise coprime. In particular there is an ideal $\mathfrak{a}$ such that

$$\langle x + \zeta y\rangle = \mathfrak{a}^p.$$

Thus $\mathfrak{a}^p$ is principal. Regularity of p means that $p \nmid h$, the class-number of $\mathbf{Q}(\zeta)$, and then Proposition 9.8(b) tells us that $\mathfrak{a}$ is principal, say $\mathfrak{a} = \langle\delta\rangle$. It follows that

$$x + \zeta y = \epsilon\delta^p$$

where ϵ is a unit.

Now we use Lemma 11.7 to conclude that

$$x + \zeta y = r\zeta^g\delta^p$$

where r is real. By Lemma 11.5 there exists $a \in \mathbf{Z}$ such that

$$\delta^p \equiv a \pmod{\mathfrak{l}^p}.$$

Hence

$$x + \zeta y \equiv ra\zeta^g \pmod{\mathfrak{l}^p}.$$

Lemma 11.3(a) shows that $\langle p \rangle \mid \mathfrak{l}^p$, so

$$x + \zeta y \equiv ra\zeta^g \pmod{\langle p \rangle}.$$

Now ζ^{-g} is a unit, so

$$\zeta^{-g}(x + \zeta y) \equiv ra \pmod{\langle p \rangle}.$$

Taking complex conjugates leads to

$$\zeta^g(x + \zeta^{-1}y) \equiv ra \pmod{\langle p \rangle}$$

and so, eliminating ra, we obtain the important congruence

$$x\zeta^{-g} + y\zeta^{1-g} - x\zeta^g - y\zeta^{g-1} \equiv 0 \pmod{\langle p \rangle}. \qquad (7)$$

Observe that $1 + \zeta$ is a unit (put $t = -1$ in equation (3) of Chapter 3). We investigate possible values for g in (7). Suppose $g \equiv 0 \pmod p$. Then $\zeta^g = 1$, the terms with x cancel, and (7) becomes

$$y(\zeta - \zeta^{-1}) \equiv 0 \pmod{\langle p \rangle}$$

and so

$$y(1 + \zeta)(1 - \zeta) \equiv 0 \pmod{\langle p \rangle}.$$

Since $1 + \zeta$ is a unit,

$$y\lambda \equiv 0 \pmod{\langle p \rangle}.$$

Now $\langle p \rangle = \langle \lambda \rangle^{p-1}$ and $p - 1 \geqslant 2$, so we have $\lambda \mid y$. Taking norms, $p \mid y$, contrary to hypothesis. Hence $g \not\equiv 0 \pmod p$. A similar argument shows that $g \not\equiv 1 \pmod p$.

We rewrite (7) in the form

$$\alpha p = x\zeta^{-g} + y\zeta^{1-g} - x\zeta^g - y\zeta^{g-1}$$

for some $\alpha \in \mathbf{Z}[\zeta]$. Note that by the previous paragraph no exponent $-g$, $1-g$, g, $g-1$ is divisible by p. We have

$$\alpha = \frac{x}{p}\zeta^{-g} + \frac{y}{p}\zeta^{1-g} - \frac{x}{p}\zeta^{g} - \frac{y}{p}\zeta^{g-1}. \qquad (8)$$

Now $\alpha \in \mathbf{Z}[\zeta]$ and $\{1, \zeta, \ldots, \zeta^{p-1}\}$ is a $\mathbf{Z}$-basis. Hence if all four exponents are incongruent modulo p we have $x/p \in \mathbf{Z}$, contrary to hypothesis. So some pair of exponents must be congruent modulo p. Since $g \not\equiv 0, 1 \pmod p$ the only possibility left is that $2g \equiv 1 \pmod p$.

But now (8) can be rewritten as

$$\begin{aligned} \alpha p \zeta^{g} &= x + y\zeta - x\zeta^{2g} - y\zeta^{2g-1} \\ &= (x-y)\lambda. \end{aligned}$$

Taking norms we get $p \mid (x-y)$, so

$$x \equiv y \pmod p.$$

By the symmetry of (5) we also have

$$y \equiv z \pmod p$$

and hence

$$0 \equiv x^p + y^p + z^p \equiv 3x^p \pmod p.$$

Since $p \nmid x$ we must have $p = 3$.

It remains to deal with the possibility $p = 3$. Note that modulo 9, cubes of numbers prime to p (namely 1, 2, 4, 5, 7, 8) are congruent either to 1 or to -1. Hence modulo 9 a solution of (5) in integers prime to 3 takes the form

$$\pm 1 \pm 1 \pm 1 \equiv 0 \pmod 9$$

which is impossible. Hence finally $p \neq 3$ and we have a contradiction. □

A complete solution of Fermat's Last Theorem (for regular primes!) is thus reduced to the case where one of x, y, or z is a multiple of p. Kummer's proof of this case also depends heavily on ideal theory and, although long, would be accessible to us at this stage, except for one fact.

We need to know that (still for p regular) if a unit in $\mathbf{Q}(\zeta)$ is congruent modulo p to a rational integer, then it is a pth power of another unit in $\mathbf{Q}(\zeta)$. The proof of this requires new methods. It seems best to refer the reader to Borevič and Šafarevič [1] pp. 378–81 for the missing details.

11.5 Regular primes

Theorem 11.8 is, of course, useless without a test for regularity. There is, in fact, quite a simple test, but once more the proofs are far beyond our present methods. We shall nonetheless sketch what is involved, and again refer the reader to Borevič and Šafarevič [1] for the details.

Everything rests on a remarkable gadget known as the *analytic class-number formula*. Let K be a number field, and define the *Dedekind zeta-function*

$$\zeta_K(x) = \sum \mathrm{N}(\mathfrak{a})^{-x}$$

where $\mathfrak{a}$ runs through all ideals of the ring of integers $\mathfrak{O}$ of K, and for the moment $1 < x < \infty$. One then proves the formula

$$\lim_{x\to 1} (x-1)\zeta_K(x) = \frac{2^{s+t}\pi^t R}{m\sqrt{|\Delta|}}\, h$$

in which s and t are the number of real, or complex conjugate pairs of, monomorphisms $K \to \mathbf{C}$; m is the number of roots of unity in K; Δ is the discriminant of K; R is a new constant called the *regulator* of K; and h is the class-number.

The point is that nearly everything on the right, except h, is quite easy to compute (R is not: it is much harder than the rest): if we could evaluate the limit on the left we could then work out h. To evaluate this limit we first extend the definition of $\zeta_K(x)$ to allow complex values of x, and then use powerful techniques from complex function theory. These involve another gadget known as a *Dirichlet L-series*.

In the case $K = \mathbf{Q}(\zeta)$ for $\zeta = e^{2\pi i/p}$, p prime, the analysis leads to an expression for h in the form of a product

$$h = h_1 h_2.$$

In this, h_2 is the class-number of the related number field $\mathbf{Q}(\zeta + \zeta^{-1})$, and h_1 is a computable integer. This would not be very helpful, except that it can be proved that *if h_1 is prime to p, then so is h_2*. (Therefore h is prime to p, or equivalently p is regular, if and only if h_1 is prime to p.)

Analysis of h_1 leads to a criterion: h_1 is divisible by p if and only if one of the numbers

$$S_k = \sum_{n=1}^{p-1} n^k \quad (k = 2, 4, \ldots, p-3)$$

is divisible by p^2.

The numbers S_k have long been associated with the *Bernouilli numbers B_k* defined by the series expansion

$$\frac{t}{e^t - 1} = 1 + \sum_{m=1}^{\infty} \frac{B_m}{m!} t^m.$$

Their values behave very irregularly: for m odd $\neq 1$ they are zero, for $m = 1$ we have $B_1 = -\frac{1}{2}$, and for even m the first few are:

$$B_2 = \frac{1}{6}, \quad B_4 = -\frac{1}{30}, \quad B_6 = \frac{1}{42},$$

$$B_8 = -\frac{1}{30}, \quad B_{10} = \frac{5}{66}, \quad B_{12} = -\frac{691}{2730},$$

$$B_{14} = \frac{7}{6}, \quad B_{16} = -\frac{3617}{510}, \ldots.$$

The connection between the S_k and the B_k may be shown to give:

Criterion 11.9. *A prime p is regular if and only if it does not divide the numerators of the Bernouilli numbers B_2, $B_4, \ldots, B_{p-3}$.* □

The first 10 irregular primes, found from this criterion,

are 37, 59, 67, 101, 103, 131, 149, 157, 233, 257. As a check, it is possible to compute the number h_1, with the following results:

p	h_1	p	h_1
3	1	43	211
5	1	47	5.139
7	1	53	4889
11	1	59	3.**59**.233
13	1	61	41.1861
17	1	67	**67**.12739
19	1	71	7^2.79241
23	3	73	89.134353
29	2^3	79	5.53.377911
31	3^2	83	3.279405653
37	**37**	89	113.118401449
41	11^2	97	577.3457.206209

Observe that h_1 is divisible by p exactly in the cases $p = 37, 59, 67$ (marked in bold type) as expected.

Exercises

11.1 If x, y, z are integers such that $x^2 + y^2 = z^2$, prove that at least one of x, y, z is a multiple of 3, at least one is a multiple of 4, and at least one is a multiple of 5.

11.2 Show that the smallest value of z for which there exist four distinct solutions to $x^2 + y^2 = z^2$ with x, y, z pairwise coprime (not counting sign changes or interchanges of x, y as distinct) is 1105, and find the four solutions.

11.3 Show that there exist no solutions in non-zero integers to the equation $x^3 + y^3 = 3z^3$. (See Hardy and Wright [4] p. 196.)

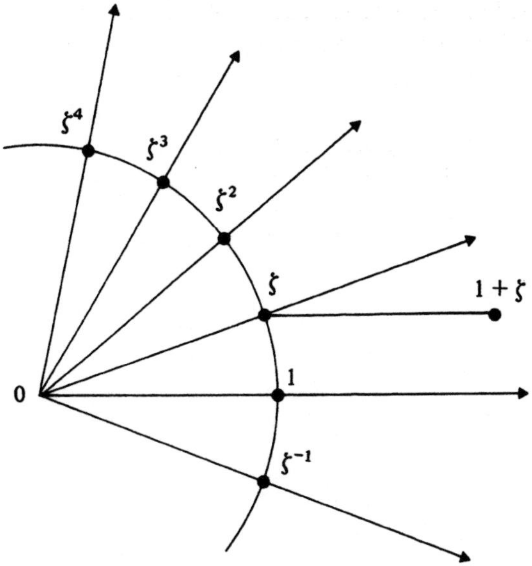

Fig. 10. Why doesn't this contradict Kummer's lemma?

11.4 Show that the general solution in rational numbers of the equation

$$x^3 + y^3 = u^3 + v^3$$

is

$$x = k(1 - (a - 3b)(a^2 + 3b^2)),$$
$$y = k((a + 3b)(a^2 + 3b^2) - 1),$$
$$u = k((a + 3b) - (a^2 + 3b^2)^2),$$
$$v = k((a^2 + 3b^2)^2 - (a - 3b)),$$

where a, b, k are rational and $k \neq 0$; or $x = y = 0$, $u = -v$; or $x = u, y = v$, or $x = v, y = u$. (*Hint*: write $x = X - Y, y = X + Y, u = U - V$, $v = U + V$, and factorize the resulting equation in $\mathbf{Q}(\sqrt{-3})$.)

11.5 For p an odd prime, show that if $\zeta = e^{2\pi i/p}$, then

$$\sqrt{\left(\frac{1-\zeta^s}{1-\zeta} \cdot \frac{1-\zeta^{-s}}{1-\zeta^{-1}}\right)}$$

is a real unit in $\mathbf{Q}(\zeta)$ for $s = 1, 2, \ldots, p-1$.

11.6 Let p be an odd prime, $\zeta = e^{2\pi i/p}$. Kummer's lemma says that the units of $\mathbf{Z}[\zeta]$, thought of in the complex plane $\mathbf{C}$, lie on equally spaced radial lines through the origin, passing through the vertices of a regular p-gon (namely the powers ζ^s). Now $1 + \zeta$ is a unit, so why does the above figure (Fig. 10, p. 209) not contradict Kummer's lemma?

CHAPTER TWELVE

Dirichlet's Units Theorem

In several instances we have seen how important it is to know something about the units in the integers of a number field. Much of the effort on Fermat's Last Theorem was concentrated in this direction. The importance of units stems from the fact that, while ideals are best suited to technicalities, there may come a point at which it is necessary to return to elements. But the generator of a principal ideal is ambiguous up to multiples by a unit. Our only constraint on the ambiguity is what we can say about units.

The most fundamental and far-reaching theorem on units is that of Dirichlet, which gives an almost complete description, in abstract terms, of the group of units in the integers of any number field. In particular it implies that this group is finitely generated. We shall prove Dirichlet's theorem in this chapter. The methods are 'geometric' in that we use Minkowski's theorem, and also a 'logarithmic' variant of the space $\mathbf{L}^{st}$.

12.1 Introduction

We have already described the units in the integers of $\mathbb{Q}(\sqrt{d})$ for negative squarefree d in Proposition 4.2. For $d = -1$ the units are $\{\pm 1, \pm i\}$, for $d = -3$ they are $\{\pm 1, \pm \omega, \pm \omega^2\}$

where $\omega = e^{2\pi i/3}$, and for all other $d < 0$, the units are just $\{\pm 1\}$.

We see that in all cases U is a finite cyclic group of even order (2, 4, or 6) whose elements are roots of unity. (It is in any case obvious that every unit of finite order is a root of unity in any number field.)

For other number fields the structure of U is more complicated. For example in $\mathbf{Q}(\sqrt{2})$ we have

$$(1 + \sqrt{2})(-1 + \sqrt{2}) = 1$$

so $\epsilon = 1 + \sqrt{2}$ is a unit. Now ϵ is not a root of unity since $|\epsilon| = 1 + \sqrt{2} \neq 1$. It follows that all the elements $\pm\, \epsilon^n$ $(n \in \mathbf{Z})$ are distinct units, so U is an infinite group. In fact, though we shall not prove it here, the $\pm\, \epsilon^n$ are all the units of $\mathbf{Q}(\sqrt{2})$; hence U is isomorphic to $\mathbf{Z}_2 \times \mathbf{Z}$.

After we have proved Dirichlet's theorem it will emerge that this more complicated structure of U is in some sense typical.

12.2 Logarithmic space

Let K be a number field of degree $n = s + 2t$, as in Chapter 8; and let $\mathbf{L}^{st}$ be as there described. We use the notation of Chapter 8 in what follows. Define a map

$$l : \mathbf{L}^{st} \to \mathbf{R}^{s+t}$$

as follows. For $x = (x_1, \ldots, x_s; x_{s+1}, \ldots, x_{s+t}) \in \mathbf{L}^{st}$ put

$$l_k(x) = \begin{cases} \log |x_k| & \text{for } k = 1, \ldots, s \\ \log |x_k|^2 & \text{for } k = s+1, \ldots, s+t. \end{cases}$$

Then set

$$l(x) = (l_1(x), \ldots, l_{s+t}(x)).$$

The additive property of the logarithm leads at once to the

$$l(xy) = l(x) + l(y) \tag{1}$$

for $x, y \in \mathbf{L}^{st}$. The set of elements of $\mathbf{L}^{st}$ with all co-ordinates non-zero forms a group under multiplication, and l defines a homomorphism of this group into $\mathbf{R}^{s+t}$. From Formula (1) of Chapter 8 it follows that

$$\sum_{k=1}^{s+t} l_k(x) = \log |\mathrm{N}(x)|. \qquad (2)$$

For $\alpha \in K$ we define

$$l(\alpha) = l(\sigma(\alpha))$$

where $\sigma : K \to \mathbf{L}^{st}$ is the standard map. This ambiguity in the use of l causes no confusion, and is tantamount to an identification of α with $\sigma(\alpha)$. Explicitly we find that

$$l(\alpha) = (\log |\sigma_1(\alpha)|, \ldots, \log |\sigma_s(\alpha)|,$$
$$\log |\sigma_{s+1}(\alpha)|^2, \ldots, \log |\sigma_{s+t}(\alpha)|^2).$$

The map $l : K \to \mathbf{R}^{s+t}$ is the *logarithmic representation* of K, and $\mathbf{R}^{s+t}$ the *logarithmic space.*

From Formula (2) of Chapter 8 and (1) above, it follows that

$$l(\alpha\beta) = l(\alpha) + l(\beta) \qquad (\alpha, \beta \in K)$$

so that l is a homomorphism from the multiplicative group $K^* = K \setminus \{0\}$ of K to the additive group of $\mathbf{R}^{s+t}$. Further we have, setting $l_k(\alpha) = l_k(\sigma(\alpha))$,

$$\sum_{k=1}^{s+t} l_k(\alpha) = \log |\mathrm{N}(\alpha)|,$$

using (3) of Chapter 8 and (2) above.

12.3 Embedding the unit group in logarithmic space

Why all these logarithms? Because the group of units is multiplicative, whereas Minkowski's theorem applies to lattices which are additive. We must pass from one milieu to the other, and it is just for this purpose that logarithms were created.

Let U be the group of units of $\mathfrak{O}$, the ring of integers of K. By restriction we obtain a homomorphism

$$l : U \to \mathbf{R}^{s+t}.$$

It is not injective, but the kernel is easily described:

Lemma 12.1. *The kernel W of $l: U \to \mathbf{R}^{s+t}$ is the set of all roots of unity belonging to U. This is a finite cyclic group of even order.*

Proof. We have $l(\alpha) = 0$ if and only if $|\sigma_i(\alpha)| = 1$ for all i. The field polynomial

$$\prod_i (t - \sigma_i(\alpha))$$

lies in $\mathbf{Z}[t]$ by Theorem 2.5a combined with Lemma 2.12. We can therefore appeal to Lemma 11.6 to conclude that all the $\sigma_i(\alpha)$ are roots of unity, in particular α itself.

The image $\sigma(\mathfrak{O})$ in $\mathbf{L}^{st}$ is a lattice by corollary 8.3, so is discrete by Theorem 6.1. Since the unit circle in $\mathbf{C}$ maps to a bounded subset in $\mathbf{L}^{st}$ it follows that $\mathfrak{O}$ contains only finitely many roots of unity, so W is finite. But any finite subgroup of K^* is cyclic (see [GT], Theorem 16.7, p. 171). Finally, W contains -1 which has order 2, so W has even order. □

Obviously the next thing to find out is the image of U in $\mathbf{R}^{s+t}$. Let us call it E. We have:

Lemma 12.2. *The image E of U in $\mathbf{R}^{s+t}$ is a lattice of dimension $\leqslant s + t - 1$.*

Proof. The norm of every unit is ± 1, so for a unit ϵ we have

$$\sum_{k=1}^{s+t} l_k(\epsilon) = \log |\mathrm{N}(\epsilon)| = \log 1 = 0.$$

Hence all points of E lie in the subspace V of $\mathbf{R}^{s+t}$ whose elements $(x_1, \ldots, x_{s+t})$ satisfy

$$x_1 + \ldots + x_{s+t} = 0.$$

This has dimension $s + t - 1$.

To prove E is a lattice it is sufficient to prove it discrete by Theorem 6.1. Let $\| \quad \|$ be the usual length function on $\mathbf{R}^{s+t}$. Suppose $0 < r \in \mathbf{R}$, and

$$\|l(\epsilon)\| < r.$$

Now $|l_k(\epsilon)| \leqslant \|l(\epsilon)\| < r$, from which we get

$$|\sigma_k(\epsilon)| < e^r \qquad (k = 1, \ldots, s)$$

$$|\sigma_{s+j}(\epsilon)|^2 < e^r \qquad (j = 1, \ldots, t).$$

Hence the set of points $\sigma(\epsilon)$ in $\mathbf{L}^{st}$ corresponding to units with $\|l(\epsilon)\| < r$ is bounded, so finite by Corollary 8.3. Hence E intersects each closed ball in $\mathbf{R}^{s+t}$ in a finite set, so is discrete.

Therefore E is a lattice. Since $E \subseteq V$ it has dimension $\leqslant s + t - 1$. □

Already we know quite a lot about U. In particular, U is finitely generated; for W is finite and $U/W \cong E$ is a lattice, so free abelian, with rank $\leqslant s + t - 1$. All that remains is to find the exact dimension of the lattice E. In fact it is $s + t - 1$, as we prove in the next section.

12.4 Dirichlet's theorem

The main thing we lack is a topological criterion for deciding whether a lattice L in a vector space V has the same dimension as V. We remedy this lack with:

Lemma 12.3. *Let L be a lattice in $\mathbf{R}^m$. Then L has dimension m if and only if there exists a bounded subset B of $\mathbf{R}^m$ such that*

$$\mathbf{R}^m = \bigcup_{x \in L} x + B.$$

Proof. If L has dimension m then we may take for B a fundamental domain of L, and appeal to Lemma 6.2.

Suppose conversely that B exists but, for a contradiction,

L has dimension $d < m$. An intuitive argument goes thus: the quotient $\mathbf{R}^m/L$ is, by Theorem 6.6, the direct product of a torus and $\mathbf{R}^{m-d}$. The condition on B says that the image of B under the natural map $\nu: \mathbf{R}^m \to \mathbf{R}^m/L$ is the whole of $\mathbf{R}^m/L$. But because B is bounded this contradicts the presence of a direct factor $\mathbf{R}^{m-d}$ which is unbounded. By taking more account of the topology than we have done hitherto, this argument can easily be made rigorous. Alternatively, we operate in $\mathbf{R}^m$ instead of $\mathbf{R}^m/L$ as follows.

Let V be the subspace of $\mathbf{R}^m$ spanned by L. If L has dimension less than m, then $\dim V < \dim \mathbf{R}^m$. Hence we can find an orthogonal complement V' to V in $\mathbf{R}^m$. The condition on B implies that $\mathbf{R}^m = \cup_{v \in V} v + B$, which means that V' is the image of B under the projection $\pi: \mathbf{R}^m \to V'$. But π is distance-preserving, so V' is bounded, an obvious contradiction. □

In fact, what we are saying topologically is that L has dimension m if and only if the quotient topological group $\mathbf{R}^m/L$ is *compact*. (This can profitably be compared with Theorem 1.13. In fact there is some kind of analogy between free abelian groups and sublattices of vector spaces; as witness to which the reader should compare Lemma 9.3 and Theorem 1.13.)

Before proving that E has dimension $s + t - 1$ it is convenient to extract one computation from the proof:

Lemma 12.4. *Let $y \in \mathbf{L}^{st}$ and let $\lambda_y : \mathbf{L}^{st} \to \mathbf{L}^{st}$ be defined by $\lambda_y(x) = yx$. Then λ_y is a linear map and*

$$\det \lambda_y = \mathrm{N}(y).$$

Proof. It is obvious that λ_y is linear. To compute $\det \lambda_y$ we use the basis (4) of Chapter 8. If

$$y = (x_1, \ldots, x_s; y_1 + iz_1, \ldots, y_t + iz_t)$$

then we obtain for $\det \lambda_y$ the expression

$$\begin{vmatrix} x_1 & & & & & & & & & \\ & \cdot & & & & & & & \mathbf{0} & \\ & & \cdot & & & & & & & \\ & & & \cdot & & & & & & \\ & & & & x_s & & & & & \\ & & & & & y_1 & -z_1 & & & \\ & & & & & z_1 & y_1 & & & \\ & & & & & & & \cdot & & \\ & \mathbf{0} & & & & & & & \cdot & \\ & & & & & & & & & \cdot \\ & & & & & & & & y_t & -z_t \\ & & & & & & & & z_t & y_t \end{vmatrix}$$

which is

$$x_1 \dots x_s(y_1^2 + z_1^2) \dots (y_t^2 + z_t^2) = \mathrm{N}(y). \qquad \square$$

The way is now clear for the proof of:

Theorem 12.5. *The image E of U in $\mathbf{R}^{s+t}$ is a lattice of dimension $s + t - 1$.*

Proof. As before let V be the subspace of $\mathbf{R}^{s+t}$ whose elements satisfy

$$x_1 + \dots + x_{s+t} = 0.$$

Then $E \subseteq V$, and $\dim V = s + t - 1$. To prove the theorem we appeal to Lemma 12.3: it is sufficient to find in V a bounded subset B such that

$$V = \bigcup_{e \in E} e + B.$$

This 'additive' property translates into a 'multiplicative' property in $\mathbf{L}^{st}$. Every point in $\mathbf{R}^{s+t}$ is the image under l of some point in $\mathbf{L}^{st}$, so every point in V is the image of some point in $\mathbf{L}^{st}$. In fact, for $x \in \mathbf{L}^{st}$, we have $l(x) \in V$ if and only if $|\mathrm{N}(x)| = 1$. So if we let

$$S = \{x \in \mathbf{L}^{st} : |\mathrm{N}(x)| = 1\}$$

then $l(S) = V$. If $X_0 \subseteq S$ is bounded, then so is $l(X_0)$, as may be verified easily. If $x \in S$ then the multiplicativity of the norm implies that $xX_0 \subseteq S$ if $X_0 \subseteq S$. In particular if ϵ is a unit then $\sigma(\epsilon)X_0 \subseteq S$. So if we can find a bounded subset X_0 of S such that

$$S = \bigcup_{\epsilon \in U} \sigma(\epsilon)X_0 \tag{3}$$

then $B = l(X_0)$ will do what is required in V.

Now we find a suitable X_0. Let M be the lattice in $\mathbf{L}^{st}$ corresponding to $\mathfrak{D}$ under σ. Consider the linear transformation $\lambda_y : \mathbf{L}^{st} \to \mathbf{L}^{st}$ ($y \in \mathbf{L}^{st}$) of Lemma 12.4. If $y \in S$ then the determinant of λ_y is $\mathrm{N}(y)$ which is ± 1. Therefore λ_y is unimodular. This, by the remark after Lemma 9.3, implies that any fundamental domain for the lattice $yM (= \lambda_y(M))$ has the same volume as a fundamental domain for M. Call this volume v.

Choose real numbers $c_i > 0$ with

$$Q = c_1 \ldots c_{s+t} > \left(\frac{4}{\pi}\right)^t v.$$

Let X be the set of $x \in \mathbf{L}^{st}$ for which

$$|x_k| < c_k \qquad (k = 1, \ldots, s)$$
$$|x_{s+j}|^2 < c_{s+j} \qquad (j = 1, \ldots, t).$$

Then by Lemma 9.2 there exists in yM a non-zero point $x \in X$. We have

$$x = y\sigma(\alpha) \qquad (0 \neq \alpha \in \mathfrak{D}).$$

Since

$$\mathrm{N}(x) = \mathrm{N}(y)\mathrm{N}(\alpha) = \pm \mathrm{N}(\alpha)$$

it follows that

$$|\mathrm{N}(\alpha)| < Q.$$

By Theorem 5.12(c) only finitely many ideals of $\mathfrak{D}$ have norm $< Q$. Considering principal ideals, and recalling that the generators of these are ambiguous up to unit multiples, it follows that there exist in $\mathfrak{O}$ only finitely many pairwise non-associated numbers

$$\alpha_1, \ldots, \alpha_N$$

whose norms are $< Q$ in absolute value. Thus for some $i = 1, \ldots, N$ we have $\alpha\epsilon = \alpha_i$, with ϵ a unit. It follows that

$$y = x\sigma(\alpha_i^{-1})\sigma(\epsilon). \tag{4}$$

Now define

$$X_0 = S \cap \left(\bigcup_{i=1}^{N} \sigma(\alpha_i^{-1})X \right). \tag{5}$$

Since X is bounded so are the sets $\sigma(\alpha_i^{-1})X$, and since N is finite X_0 is bounded. Obviously X_0 does not depend on the choice of $y \in S$.

But now, since y and $\sigma(\epsilon) \in S$, we have $x\sigma(\alpha_i^{-1}) \in S$, hence $x\sigma(\alpha_i^{-1}) \in X_0$. Then (4) shows that

$$y \in \sigma(\epsilon)X_0.$$

Hence (3) holds, y being an arbitrary element of S, and the theorem is proved. □

We can easily put this result into a more explicit form, obtaining the *Dirichlet Units Theorem*:

Theorem 12.6. *The group of units of* $\mathfrak{O}$ *is isomorphic to*

$$W \times \mathbf{Z} \times \ldots \times \mathbf{Z}$$

where W *is as described in Lemma* 12.1 *and there are* $s + t - 1$ *direct factors* $\mathbf{Z}$.

Proof. By Theorem 12.5, $U/W \cong \mathbf{Z} \times \ldots \times \mathbf{Z} = \mathbf{Z}^{s+t-1}$. Since W is finite it follows that U is a finitely generated abelian group, hence a direct product of cyclic groups (see [HH], p. 153). Since W is finite and U/W torsion-free it follows that W is the set of elements of U of finite order, which is the product of all the finite cyclic factors in the direct decomposition. The other factors are all infinite cyclic; looking at U/W tells us there are exactly $s + t - 1$ of them. □

In more classical terms, Dirichlet's theorem asserts the existence of a system of $s + t - 1$ *fundamental units*

$$\eta_1, \ldots, \eta_{s+t-1}$$

such that every unit of $\mathfrak{O}$ is representable *uniquely* in the form

$$\zeta \cdot \eta_1^{r_1} \ldots \eta_{s+t-1}^{r_{s+t-1}}$$

for a root of unity ζ and rational integers r_i.

Let us return briefly to $\mathbf{Q}(\sqrt{2})$, which we looked at in Section 1 of this chapter. For this field, $s = 2$, $t = 0$, so $s + t - 1 = 1$. Hence U is of the form $W \times \mathbf{Z}$ where W consists of the roots of unity in $\mathbf{Q}(\sqrt{2})$. These are just ± 1, so we get $U \cong \mathbf{Z}_2 \times \mathbf{Z}$ as asserted in Section 1. Note, however, that we have still not proved that $1 + \sqrt{2}$ is a fundamental unit. In fact, this is true in general of Dirichlet's theorem: it does not allow us to find any *specific* system of fundamental units. Other methods can be developed to solve this problem, and the Dirichlet theorem is still needed to tell us when we have found sufficiently many units.

Exercises

12.1 Find units, not equal to 1, in the rings of integers of the fields $\mathbf{Q}(\sqrt{d})$ for $d = 2, 3, 5, 6, 7, 10$.

12.2 Use Dirichlet's theorem to prove that for any square-free positive integer d there exist infinitely many integer solutions x, y to the *Pell equation*

$$x^2 - dy^2 = 1.$$

(Really this should not be called the Pell equation, since Pell did not solve it. It was mistakenly attributed to him by Euler, and the name stuck.)

12.3 Prove that $1 + \sqrt{2}$ is a fundamental unit for $\mathbf{Q}(\sqrt{2})$.

12.4 Let $\eta_1, \ldots, \eta_{s+t-1}$ be a system of fundamental units for a number field K. Show that the *regulator*

$$R = |\det(\log|\sigma_i(\eta_j)|)|$$

is independent of the choice of $\eta_1, \ldots, \eta_{s+t-1}$.

12.5 Show that the group of units of a number field K is finite if and only if $K = \mathbf{Q}$, or K is an imaginary quadratic field.

12.6 Show that a number field of odd degree contains only two roots of unity.

APPENDIX

Quadratic residues

The topic of quadratic residues has been segregated because it is independent of the rest of the book; and, since the methods used are computational (and on the whole straightforward), it may in fact be read at any time. Suitable places in a regular schedule of work might be at the very beginning, or alongside Chapter 3 (which is fairly short) to give a block of work commensurate in difficulty with the other chapters in the first part of the book. The appendix is laid out in the same manner as the main part of the text, ending with a set of exercises.

An integer k which is prime to a positive integer m is said to be a *quadratic residue modulo m* if there exists $z \in \mathbf{Z}$ such that

$$z^2 \equiv k \pmod{m}.$$

Denoting the residue class of k modulo m by $\bar{k}$, this can be rephrased as: $\bar{k}$ is both a unit and a square in $\mathbf{Z}_m$.

We shall investigate the question of quadratic residues by determining the structure of the units in $\mathbf{Z}_m$. We shall also show how knowledge of quadratic residues solves the more general question of finding solutions of the quadratic equation

$$ax^2 + bx + c = 0$$

in $\mathbf{Z}_m$.

The most remarkable theorem about quadratic residues is known as the *quadratic reciprocity law* which states:

If p, q are distinct odd primes, at least one of which is congruent to 1 modulo 4, then p is a quadratic residue of q if and only if q is a quadratic residue of p; otherwise precisely one of p, q is a quadratic residue of the other.

The reciprocal nature of the relationship between p and q in the first case gives rise to the name of the law. Gauss first proved it in 1796 when he was but eighteen years old. The result had been conjectured earlier by Euler and Legendre, though Gauss said that he did not know this at the time. He thought so highly of it that he called it 'the gem of higher arithmetic' and developed six different proofs. In the 19th century it continued to arouse interest, with more that fifty different methods of proof from such as Cauchy, Eisenstein, Jacobi, Kronecker, Kummer, Liouville and Zeller. In fact it was when Kummer was studying this and certain higher reciprocity laws that he came upon his partial proof of Fermat's Last Theorem. In 1850 Kummer referred to the higher reciprocity laws as the 'the pinnacle of contemporary number theory', regarding Fermat's Last Theorem as a 'curiosity' (see Edwards [23]). It is only right and proper, therefore, that any text on Fermat's Last Theorem should include a description of the result that so fascinated number-theorists and whose study led to Kummer's proof of Fermat's Theorem as a by-product.

A.1. Quadratic equations in $\mathbf{Z}_m$

An obvious topic of number-theoretic study is the solution of polynomial equations modulo a positive integer m.

A linear equation

$$ax + b \equiv 0 \pmod{m} \tag{1}$$

can clearly be solved when a is prime to m, because there exist integers c, d such that

$$ac + dm = 1.$$

Whence multiplying (1) by c and simplifying gives

$$x \equiv -bc \qquad (\text{mod } m).$$

If, on the other hand, a, m have a highest common factor h, then (1) can only have a solution if h divides b, in which case, writing $a = a_0h$, $b = b_0h$, $m = m_0h$, we can reduce (1) to

$$a_0x + b_0 \equiv 0 \qquad (\text{mod } m_0)$$

where once more a_0, m_0 are coprime. Thus the solution of linear equations modulo m is straightforward.

Quadratic equations are more interesting. The equation

$$ax^2 + bx + c \equiv 0 \qquad (\text{mod } m) \qquad (2)$$

where m does not divide a may be simplified by multiplying throughout (including the modulus) by $4a$ and completing the square to get the equivalent equation

$$4a^2x^2 + 4abx + 4ac \equiv 0 \qquad (\text{mod } 4am)$$

or

$$(2ax + b)^2 \equiv b^2 - 4ac \qquad (\text{mod } 4am).$$

Now substituting $4am = m_0$, $2ax + b \equiv z \text{ mod } m_0$ and $b^2 - 4ac \equiv k \text{ mod } m_0$, we replace (2) by the two equations:

$$z^2 \equiv k \qquad (\text{mod } m_0) \qquad (3)$$

$$2ax + b \equiv z \qquad (\text{mod } m_0). \qquad (4)$$

If we find z from (3), we can then attack (4) by the given method for linear congruences, so the solution of the general quadratic (2) reduces to solving (3). We can further reduce this if k, m_0 are not coprime by supposing they have highest common factor h where $k = k_0h$ and $m_0 = m_1h$, and then factorizing h as

$$h = e^2r$$

where e^2 is the largest square factor of h. Then for (3) to have a solution for z we must have er as a factor of z. Let $z = erw$; then

$$e^2r^2w^2 \equiv k_0e^2r \qquad (\text{mod } m_1e^2r)$$

so

$$rw^2 \equiv k_0 \qquad (\text{mod } m_1). \qquad (5)$$

Now suppose that the highest common factor of r, m_1 is s. For (5) to have a solution we must have s as a factor of k_0. But k_0, m_1 are coprime, so $s = 1$ and r, m_1 are also coprime. Thus there exists an integer d prime to m_1 such that

$$dr \equiv 1 \pmod{m_1},$$

so multiplying (5) by d and simplifying gives

$$w^2 \equiv dk_0 \pmod{m_1}. \tag{6}$$

But d and k_0 are both prime to m_1, so putting $dk_0 = k_1$, the solution of the general quadratic equation (2) reduces to linear congruences and

$$w^2 \equiv k_1 \pmod{m_1}$$

where k_1, m_1 are coprime.

If $m > 1$ and k are integers, recall that k is a quadratic residue modulo m if

(i) k, m are coprime

(ii) There exists $w \in \mathbf{Z}$ such that

$$w^2 \equiv k \pmod{m}.$$

If $\bar{k}$ is the residue class of k in $\mathbf{Z}_m$, then these conditions are equivalent to

(i)′ $\bar{k}$ is a unit in $\mathbf{Z}_m$,

(ii)′ $\bar{w}^2 = \bar{k}$ in $\mathbf{Z}_m$,

so that $\bar{k}$ is both a unit and a square in $\mathbf{Z}_m$. We shall attack the problem of finding quadratic residues by first computing the structure of the units in $\mathbf{Z}_m$.

A.2 The units of $\mathbf{Z}_m$

An element $\bar{k} \in \mathbf{Z}_m$ is a unit if and only if k, m are coprime, so the units $U(\mathbf{Z}_m)$ of $\mathbf{Z}_m$ are given by

$$U(\mathbf{Z}_m) = \{\bar{k} \in \mathbf{Z}_m \mid 1 \leqslant k < m;\ k, m \text{ coprime}\}.$$

The number of elements in $U(\mathbf{Z}_m)$ is called the *Euler function* $\phi(m)$ and is equal to the number of positive integers k less than m and prime to it. For example $\phi(10) = 4$ since

1, 3, 7, 9 are the integers between 1 and 10 and prime to 10, and there are four of them. For later reference we record:

Lemma A.1. *If p is prime, then $\phi(p^e) = p^{e-1}(p-1)$, and in particular $\phi(p) = p-1$.*

Proof. There are $p^e - 1$ elements satisfying $1 \leqslant k < p^e$, and of these, if $k = rp$, then

$$1 \leqslant rp < p^e$$

implies

$$1 \leqslant r < p^{e-1},$$

so there are $p^{e-1} - 1$ elements not prime to p, giving

$$\phi(p^e) = (p^e - 1) - (p^{e-1} - 1). \qquad \square$$

The units $U(\mathbf{Z}_m)$ form a group under multiplication whose structure we shall compute. First we factorize $m = p_1^{e_1} \ldots p_r^{e_r}$ where $p_1, \ldots, p_r$ are distinct primes and reduce the problem to considering each prime separately. For typographical reasons we write $p_i^{e_i} = P_i$.

Lemma A.2. *If $m = p_1^{e_1} \ldots p_r^{e_r}$ where $p_1, \ldots, p_r$ are distinct primes, then, writing $p_i^{e_i} = P_i$, there is a ring isomorphism*

$$\mathbf{Z}_m \cong \mathbf{Z}_{P_1} \times \ldots \times \mathbf{Z}_{P_r}.$$

Proof. Define $\pi : \mathbf{Z} \to \mathbf{Z}_{P_1} \times \ldots \times \mathbf{Z}_{P_r}$ by $\pi(k) = (k_1, \ldots, k_r)$ where k_i is the residue class of k modulo $P_i = p_i^{e_i}$. Clearly π is a ring homomorphism and $k \in \ker \pi$ if and only if P_i divides k for every i, which implies $\ker \pi = \langle m \rangle$ (the ideal generated by m). Thus π induces a monomorphism

$$\bar{\pi} : \mathbf{Z}_m \to \mathbf{Z}_{P_1} \times \ldots \times \mathbf{Z}_{P_r}.$$

But $\mathbf{Z}_{P_1} \times \ldots \times \mathbf{Z}_{P_r}$ has $P_1 \times \ldots \times P_r = m$ elements, so $\bar{\pi}$ is in fact an isomorphism. $\square$

Lemma A.3. *With the above notation,*

$$U(\mathbf{Z}_m) \cong U(\mathbf{Z}_{P_1}) \times \ldots \times U(\mathbf{Z}_{P_r}).$$

Proof. Under the ring isomorphism $\bar{\pi}$, an element $\bar{k} \in \mathbf{Z}_m$ is a unit if and only if $\bar{\pi}(k)$ is a unit, which holds if and only if each of its components is a unit. □

Lemma A.3 reduces the study of units in $\mathbf{Z}_m$ to the case of $\mathbf{Z}_{p^e}$ where p is prime. To tackle this case we begin with $e = 1$. Here we shall see that $U(\mathbf{Z}_p)$ is a cyclic group of order $p - 1$. It might seem that the easiest way to show this would be to find a generator, but to give an explicit construction for one proves to be moderately intricate. So we attack the problem indirectly by introducing an auxiliary notion which will prove useful in several ways:

The *exponent* h of a finite group G is defined to be the smallest positive integer such that (in multiplicative notation) $x^h = 1$ for all $x \in G$. By Lagrange's Theorem, $x^n = 1$ where n is the order of G, so clearly $h \leqslant n$. Also, for any $x \in G$, if x has order k, then k divides h. An alternative definition of the exponent is, therefore, the least common multiple of all the orders of the elements in G.

We intend to establish that in an abelian group (such as $U(\mathbf{Z}_p)$) there actually exists an element x_0 of order h. This will then establish two facts; first that (by Lagrange's Theorem) the exponent h divides the order of G; and second, if we can demonstrate that the exponent *equals* the order, then G is cyclic with generator x_0.

To demonstrate the existence of such an x_0, we begin with:

Lemma A.4. *In an abelian group G, if a, b have finite orders q, r which are coprime, then ab has order qr.*

Proof. $(ab)^{qr} = a^{qr}b^{qr} = 1$. Now suppose $(ab)^s = 1$, then

$$a^s = b^{-s}$$

so the elements a^s and b^{-s} must have the same order k; however the order of a^s divides the order of a, so k divides q, similarly k divides r. Since q, r are coprime, $k = 1$, which implies

$$a^s = b^{-s} = 1,$$

whence q divides s, r divides s, and since q, r are coprime, qr divides s, completing the proof. □

Lemma A.5. *If the finite abelian group G has exponent h, then there exists $x_0 \in G$ such that the order of x_0 is h.*

Proof. Let p^k be the highest power of a prime p dividing h. It is easy to see that G must have an element x of order $p^k r$ where r is prime to p (for if not, the highest power of p dividing the order of every element is p^{k-1} or less, contrary to the definition of h). The element $y = x^r$ is then of order p^k. Find elements y for all distinct primes p dividing h and then use Lemma A.4. □

Proposition A.6. *$U(\mathbf{Z}_p)$ is a cyclic group of order $p - 1$.*

Proof. The order of $U(\mathbf{Z}_p)$ is $\phi(p) = p - 1$ by Lemma A.1. To show $U(\mathbf{Z}_p)$ is cyclic by Lemma A.5, we need only verify that the exponent h of $U(\mathbf{Z}_p)$ is $p - 1$. Certainly $h \leqslant p - 1$. Conversely, every element of $U(\mathbf{Z}_p)$ satisfies $x^h = 1$ by definition, and interpreting $x^h - 1 = 0$ as a polynomial equation over $\mathbf{Z}_p$, this has, at most, h roots, hence $p - 1 \leqslant h$. □

Examples. $U(\mathbf{Z}_2) = \{\bar{1}\}$, generator $\bar{1}$,
$U(\mathbf{Z}_3) = \{\bar{1}, \bar{2}\}$, generator $\bar{2}$,
$U(\mathbf{Z}_5) = \{\bar{1}, \bar{2}, \bar{3}, \bar{4}\}$, generators $\bar{2}, \bar{3}$,
$U(\mathbf{Z}_7) = \{\bar{1}, \bar{2}, \bar{3}, \bar{4}, \bar{5}, \bar{6}\}$, generators $\bar{3}, \bar{5}$.

If $\bar{s}$ is a generator of $U(\mathbf{Z}_p)$, then $s \in \mathbf{Z}$ is called a *primitive root modulo p*. Such primitive roots will play a central part in our computations, since every element of $U(\mathbf{Z}_p)$ is of the form $\bar{s}^r$ where s is a primitive root modulo p. If we can find a primitive root, then because the order of $U(\mathbf{Z}_p)$ is even, the even powers $\bar{s}^{2r}$ are clearly quadratic residues and the odd powers $\bar{s}^{2r+1}$ are not. In general we shall not attack the problem this way (because we do not know the value of s), but it illustrates the theoretical importance of primitive roots. We isolate two properties which will prove essential later:

Lemma A.7. (i) *If s is a primitive root modulo p, then so is s^r if and only if $r, p-1$ are coprime.*

(ii) *If s is a primitive root modulo p and k is a positive integer, then there is another primitive root λ modulo p such that $\bar{s} = \bar{\lambda}^{p^k}$*

Proof. (i) If $r, p-1$ are coprime, then there exist integers a, b such that $ar + b(p-1) = 1$, so

$$\bar{s} = \bar{s}^{ar+b(p-1)} = (\bar{s}^r)^a,$$

and $\bar{s}^r$ generates $U(\mathbf{Z}_p)$.

Conversely, if $r, p-1$ were to have common factor $d > 1$, where $p-1 = dq$, $r = dc$, then

$$(\bar{s}^r)^q = \bar{s}^{dcq} = \bar{s}^{(p-1)c} = 1,$$

so $\bar{s}^r$ cannot generate $U(\mathbf{Z}_p)$.

(ii) Since $p^k, p-1$ are coprime, there exist integers a, b where a is prime to $p-1$ such that

$$ap^k + b(p-1) = 1.$$

Hence $\lambda = s^a$ is a primitive root modulo p by part (i) and

$$\bar{\lambda}^{p^k} = \bar{s}^{ap^k} = \bar{s}^{ap^k+b(p-1)} = \bar{s}. \qquad \square$$

We are now in a position to describe the structure of $U(\mathbf{Z}_{p^e})$ for prime p, which we shall do by explicit computation, first for an odd prime.

Proposition A.8. *If p is a prime, $p \neq 2$, $e \geqslant 2$, then $U(\mathbf{Z}_{p^e})$ is a cyclic group of order $p^{e-1}(p-1)$ with generator $\bar{s}(\overline{1+p})$, where $\bar{s}$ is a primitive root modulo p.*

Proof. Since the order of $U(\mathbf{Z}_{p^e})$ is $p^{e-1}(p-1)$ by Lemma A.1, and $p-1, p^{e-1}$ are coprime, then using Lemma A.4 it is sufficient to show $\bar{s}$ has order $p-1$ and $\overline{1+p}$ has order p^{e-1} in $U(\mathbf{Z}_{p^e})$. In the case of $\bar{s}$ we have

$$\bar{s}^{p-1} = \bar{\lambda}^{p^{e-1}(p-1)} \qquad \text{using Lemma A.7 (ii)}$$

$$= 1 \pmod{p^e} \text{ by Lagrange's theorem.}$$

On the other hand

$$s^r \equiv 1 \pmod{p^e}$$

implies

$$s^r \equiv 1 \pmod{p}$$

and since s is a primitive root modulo p we have

$$s^r \not\equiv 1 \pmod{p^e} \qquad \text{for } 1 < r < p-1,$$

demonstrating that $\bar{s}$ is of order $p-1$ in $U(\mathbf{Z}_{p^e})$. To prove that $\overline{1+p}$ is of order p^{e-1}, we establish by induction on $e \geqslant 2$ that

$$(1+p)^{p^{e-2}} \equiv 1 + kp^{e-1} \pmod{p^e} \tag{7}$$

where k depends on e, but $k \not\equiv 0 \pmod{p}$. For $e = 2$ this is true with $k = 1$. Assume it true for some $e \geqslant 2$. Then

$$\begin{aligned}(1+p)^{p^{e-2}} &= 1 + kp^{e-1} + rp^e \\ &= 1 + sp^{e-1}\end{aligned}$$

where $s = k + rp \not\equiv 0 \pmod{p}$.
Hence

$$\begin{aligned}(1+p)^{p^{e-1}} &= (1 + sp^{e-1})^p \\ &= 1 + psp^{e-1} + \binom{p}{2} s^2 p^{2(e-1)} + \ldots + s^p p^{p(e-1)}.\end{aligned}$$

For $e \geqslant 2$ and prime $p \neq 2$ this is of the form

$$1 + sp^e + bp^{e+1}.$$

(For $p = 2$ this only breaks down when $e = 2$.) Hence

$$(1+p)^{p^{e-1}} \equiv 1 + sp^e \pmod{p^{e+1}} \tag{8}$$

where $s \not\equiv 0 \pmod{p}$, completing the induction proof of (7).
Then (7) implies

$$(\overline{1+p})^{p^{e-2}} \neq \bar{1} \text{ in } \mathbf{Z}_{p^e}$$

and (8) implies

$$(\overline{1+p})^{p^{e-1}} = \bar{1} \text{ in } \mathbf{Z}_{p^e},$$

which together show $\overline{1+p}$ is of order p^{e-1} in $U(\mathbf{Z}_{p^e})$. □

Since the proof in Proposition 8 breaks down for $p = 2$, we must treat this case separately. We find:

Proposition A.9. $U(\mathbf{Z}_4) = \{\bar{1}, -\bar{1}\}$ *is cyclic with generator* $-\bar{1}$. *For* $e \geqslant 3$, $U(\mathbf{Z}_{2^e})$ *is not cyclic, but* $-\bar{1}$ *is of order* 2, $\bar{5}$ *is of order* 2^{e-2} *and* $U(\mathbf{Z}_{2^e})$ *is the direct product of the cyclic groups generated by* $-\bar{1}, \bar{5}$.

Proof. The assertion concerning $U(\mathbf{Z}_4)$ is trivial. For $e \geqslant 3$, by Lemma A.1, the order of $U(\mathbf{Z}_{2^e})$ is 2^{e-1}. Clearly the order of $-\bar{1}$ in $U(\mathbf{Z}_{2^e})$ is 2. For the element $\bar{5}$ we note that by induction on $e \geqslant 3$ we may establish

$$5^{2^{e-3}} = (1 + 2^2)^{2^{e-3}} \equiv 1 + 2^{e-1} \qquad (\text{mod } 2^e).$$

Hence $\bar{5}^{2^{e-3}} \neq \bar{1}$ in $\mathbf{Z}_{2^e}$, but

$$5^{2^{e-2}} \equiv (1 + 2^{e-1})^2 \equiv 1 \qquad (\text{mod } 2^e)$$

which demonstrates that $\bar{5}$ is of order 2^{e-2} in $U(\mathbf{Z}_{2^e})$. Now $-\bar{1}$ is not a power of $\bar{5}$ in $U(\mathbf{Z}_{2^e})$ since

$$-1 \not\equiv 5^r \qquad (\text{mod } 4),$$

so certainly

$$-1 \not\equiv 5^r \qquad (\text{mod } 2^e).$$

Hence if C is the cyclic subgroup generated by $\bar{5}$, the cosets $C, -\bar{1}C$ are disjoint. But the index of C in $U(\mathbf{Z}_{2^e})$ is $2^{e-1}/2^{e-2} = 2$, so these two cosets exhaust $U(\mathbf{Z}_{2^e})$. Thus every element of $U(\mathbf{Z}_{2^e})$ is uniquely of the form $(-\bar{1})^a\bar{5}^b$ where $a = 0$ or 1 and $0 \leqslant b < 2^{e-2}$. Since multiplication is commutative, $U(\mathbf{Z}_{2^e})$ is the direct product of the cyclic subgroups generated by $-\bar{1}, \bar{5}$. The exponent of $U(\mathbf{Z}_{2^e})$ is 2^{e-2} which is less than the order, so $U(\mathbf{Z}_{2^e})$ cannot be cyclic. □

Having described the structure of $U(\mathbf{Z}_{p^e})$, we are now in a position to investigate quadratic residues.

A.3 Quadratic residues

As in the last section we can speedily reduce the problem of finding residues modulo m to the case of prime powers:

Proposition A.10. *If $m = p_1^{e_1} \ldots p_r^{e_r}$ where $p_1, \ldots, p_r$ are distinct primes and k is relatively prime to m, then k is a quadratic residue modulo m if and only if it is a quadratic residue of $p_i^{e_i}$ for $1 \leqslant i \leqslant r$.*

Proof. Using the isomorphism $\bar{\pi}: U(\mathbf{Z}_m) \to U(\mathbf{Z}_{P_1}) \times \ldots \times U(\mathbf{Z}_{P_r})$ of Lemma A.3, $\bar{k}$ is a square if and only if each component of $\bar{\pi}(\bar{k})$ is a square, and the ith component is the residue class of k modulo $p_i^{e_i}$. □

This reduces the general problem of finding quadratic residues to the simpler problem of finding quadratic residues modulo a prime power. Following the last section we distinguish between the case of an odd prime and $p = 2$ first, because this can be given an immediate answer:

Proposition A.11. *The odd integer k is a quadratic residue modulo 4 if and only if $k \equiv 1 \pmod 4$, and is a quadratic residue modulo 2^e for $e \geqslant 3$ if and only if $k \equiv 1 \pmod 8$.*

Proof. Since $U(\mathbf{Z}_4) = \{\bar{1}, \bar{3}\}$ the only square in $U(\mathbf{Z}_4)$ is $\bar{1}$. For $e \geqslant 3$, if $\bar{z}^2 = \bar{k}$ in $U(\mathbf{Z}_{2^e})$, we use proposition 9 to write

$$\bar{z} = (-\bar{1})^a \bar{5}^b, \qquad \bar{k} = (-\bar{1})^c \bar{5}^d,$$

then

$$(-\bar{1})^{2a} \bar{5}^{2b} = (-\bar{1})^c \bar{5}^d,$$

whence c is even and $2b \equiv d \pmod{2^{e-2}}$. Given d, the congruence can be solved for b if and only if d is even. Thus $\bar{k}$ is a quadratic residue modulo 2^e if and only if

$$\bar{k} = \bar{5}^d$$

in $\mathbf{Z}_{2^e}$ where d is even. Putting $d = 2r$ this implies

$$k \equiv 5^{2r} \pmod{2^e}$$

for $e \geqslant 3$, hence

$$k \equiv 25^r \pmod 8$$
$$\equiv 1 \pmod 8.$$

Conversely if $k \equiv 1 \pmod 8$ and $k \equiv (-1)^c 5^d \pmod{2^e}$, then

$$(-1)^c 5^d \equiv 1 \qquad (\text{mod } 8).$$

This can only happen when c, d are even and then $(-\bar{1})^c \bar{5}^d$ is a square in $U(\mathbf{Z}_{2^e})$. □

In the case p odd we first characterize the quadratic residues modulo p by using a primitive root modulo p:

Lemma A.12. *If s is a primitive root modulo p, then $\bar{k} = \bar{s}^a$ is a quadratic residue if and only if a is even.*

Proof. If $a = 2b$, then $\bar{k} = (\bar{s}^b)^2$. Now $\bar{s}$ has even order $p - 1$, it cannot be a square, nor can $\bar{s}^a$ for a odd. □

This characterization of quadratic residues modulo p immediately gives us:

Proposition A.13. *If p is an odd prime, then k is a quadratic residue modulo p^e for $e \geqslant 2$ if and only if k is a quadratic residue modulo p.*

Proof. If $z^2 \equiv k$ (mod p^e), then clearly $z^2 \equiv k$ (mod p), so a quadratic residue modulo p^e also serves modulo p. Conversely, suppose k is a quadratic residue modulo p. By Proposition A.8 we can write $k \equiv s^a(1 + p)^a$ (mod p^e), and reducing this modulo p gives $k \equiv s^a$ (mod p), so Lemma A.12 tells us $a = 2b$ for an integer b, whence $k \equiv [s^b(1 + p)^b]^2$ (mod p^e) and k is a quadratic residue modulo p^e. □

This leaves us with the central core of the problem: determining quadratic residues modulo an odd prime p. Legendre, who published two volumes on number theory in 1830, introduced a deceptively simple notation which is ideally suited to the task. He defined the symbol (k/p) for an odd prime p and an integer k not divisible by p as

$$(k/p) = \begin{cases} +1 & \text{if } k \text{ is a quadratic residue modulo } p \\ -1 & \text{otherwise.} \end{cases}$$

The value of this notation can be seen by writing $\bar{k} = \bar{s}^a$ where s is a primitive root modulo p. By Lemma A.12, k is a quadratic residue modulo p if and only if a is even, hence

$$(k/p) = (-1)^a.$$

From this it is easy to deduce the following useful properties:

Proposition A.14. (i) $k \equiv r \pmod{p}$ *implies* $(k/p) = (r/p)$,
(ii) $(kr/p) = (k/p)(r/p)$.

Proof. (i) is immediate, and (ii) follows by writing $\bar{k} = \bar{s}^a$, $\bar{r} = \bar{s}^b$, whence $\overline{kr} = \bar{s}^{a+b}$ and

$$(kr/p) = (-1)^{a+b} = (-1)^a(-1)^b = (k/p)(r/p). \qquad \square$$

It is now possible to give a computational test for quadratic residues:

Proposition A.15 (Euler's criterion). *For an odd prime p and an integer k not divisible by p,*

$$(k/p) \equiv k^{(p-1)/2} \pmod{p}.$$

Proof. For a primitive root s modulo p we have $s^{p-1} \equiv 1 \pmod{p}$, and since $p-1$ is even,

$$(s^{(p-1)/2} - 1)(s^{(p-1)/2} + 1) = (s^{p-1} - 1) \equiv 0 \pmod{p}.$$

Because $s^{(p-1)/2} \not\equiv 1 \bmod p$, we deduce

$$s^{(p-1)/2} \equiv -1 \pmod{p}.$$

Hence, writing $\bar{k} = \bar{s}^a$ as before, we have

$$\begin{aligned} (k/p) &= (-1)^a \\ &\equiv (s^{(p-1)/2})^a \pmod{p} \\ &= (s^a)^{(p-1)/2} \\ &\equiv k^{(p-1)/2} \pmod{p}. \qquad \square \end{aligned}$$

Example. k is a quadratic residue mod 5 if $k^2 \equiv 1 \pmod 5$, giving $k = 1, 4$.

We soon see the weakness in this criterion if we attempt to find the quadratic residues modulo a larger prime, for example $p = 19$. In this case k is a quadratic residue if and only if $k^9 \equiv 1 \pmod{19}$, and the calculations concerned involve more work than just calculating all the squares of elements in $U(\mathbf{Z}_{19})$ and solving the problem by inspection. However, Euler's criterion can be used to deduce a much more useful test, due to Gauss.

What Gauss did was to partition the units modulo p by writing them in the form

$$\begin{aligned} U(\mathbf{Z}_p) &= \{-\overline{(p-1)/2}, \ldots, -\bar{2}, -\bar{1}, \bar{1}, \bar{2}, \ldots, \overline{(p-1)/2}\} \\ &= N \cup P \end{aligned}$$

where

$$N = \{-\overline{(p-1)/2}, \ldots, -\bar{2}, -\bar{1}\}$$

and

$$P = \{\bar{1}, \bar{2}, \ldots, \overline{(p-1)/2}\}.$$

For instance, if $p = 7$, then

$$N = \{-\bar{3}, -\bar{2}, -\bar{1}\}, \qquad P = \{\bar{1}, \bar{2}, \bar{3}\}.$$

Using the usual multiplicative notation $aS = \{as \mid s \in S\}$, we can write $N = (-\bar{1})P$. To find out if k is a quadratic residue, Gauss computed the set $\bar{k}P$ and proved:

Proposition A.16 (Gauss' criterion). *With the above notation, if $\bar{k}P \cap N$ has ν elements, then $(k/p) = (-1)^\nu$.*

Proof. Since $\bar{k}$ is a unit, the elements of $\bar{k}P$ are distinct and so $|\bar{k}P| = |P|$. Furthermore if $\bar{a}, \bar{b}$ are distinct elements of P, then we may take $0 < a < b \leqslant (p-1)/2$. We cannot have $\bar{k}\bar{a} = \bar{r}$ and $\bar{k}\bar{b} = -\bar{r}$, for that implies $k(a+b)$ is divisible by p, hence $a + b$ is divisible by p, contradicting the inequalities satisfied by a, b. Thus the elements $\bar{k}, \bar{k}\bar{2}, \ldots, \bar{k}\,\overline{(p-1)/2}$ of $\bar{k}P$ consist precisely of the elements $\pm\bar{1}, \pm\bar{2}, \ldots, \pm\overline{(p-1)/2}$, possibly in a different order, where the number of minus signs is the number of elements of $\overline{k}P$ in N. Hence

$$\overline{k}.\overline{k}\overline{2}\ldots\overline{k}\,\overline{(p-1)/2} = (\pm\overline{1})(\pm\overline{2})\ldots(\pm\overline{(p-1)/2})$$

so

$$\overline{k}^{(p-1)/2} = (-1)^{\nu}$$

where ν is the number of elements in $\overline{k}P \cap N$.
Thus

$$k^{(p-1)/2} \equiv (-1)^{\nu} \qquad (\text{mod } p)$$

and Euler's criterion gives

$$(k/p) = (-1)^{\nu}. \qquad \square$$

Example. Is 3 a quadratic residue modulo 19? To answer this we calculate $\overline{3}P = \{3, 6, 9, 12, 15, 18, 2, 5, 8\}$

$$= \{3, 6, 9, 2, 5, 8\} \cup \{-7, -4, -1\},$$

so $\nu = 3$ and Gauss' criterion tells us that 3 is not a quadratic residue modulo 19.

These two criteria take us further in the search for quadratic residues k modulo an odd prime p, for by factorizing

$$k = (-1)^a 2^b p_1^{e_1} \ldots p_r^{e_r}$$

then k is certainly a square if $a, b, e_1, \ldots, e_r$ are even and will be a quadratic residue if the factors with odd exponent are quadratic residues. Thus the question of quadratic residues is finally reduced to the question of determining whether -1, 2 or an odd prime q (distinct from p) are quadratic residues modulo an odd prime p. The given criteria solve the question for -1, 2:

Proposition A.17. (i) $(-1/p) = (-1)^{(p-1)/2}$, *so* -1 *is a quadratic residue modulo p if and only if* $p \equiv 1$ (mod 4).

(ii) $(2/p) = (-1)^{(p^2-1)/8}$, *so* 2 *is a quadratic residue modulo p if and only if* $p \equiv \pm 1$ (mod 8).

Proof. (i) is a trivial consequence of Euler's criterion.
(ii) $2P = \overline{2}, \overline{4}, \ldots, \overline{p-1}$, so $|\overline{2}P \cap N|$ is $\nu = (p-1)/2 - r$ where r is the largest integer such that $2r \leqslant (p-1)/2$.

Case 1. $(p-1)/2$ is even and $2r=(p-1)/2$, whence $\nu=(p-1)/2-(p-1)/4=(p-1)/4$. Thus $(2/p)=(-1)^{(p-1)/4}$.
Case 2. $(p-1)/2$ is odd and $2r=(p-1)/2-1$, so that $\nu=(p-1)/2-(p-1)/4+\frac{1}{2}=(p+1)/4$. Thus $(2/p)=(-1)^{(p+1)/4}$. We can put these two cases together by noting that in the first case $(p-1)/2$ is even if and only if $(p+1)/2$ is odd. Raising $(-1)^n$ to an odd power does not change it, so in case 1

$$(2/p) = [(-1)^{(p-1)/4}]^{(p+1)/2} = (-1)^{(p^2-1)/8}.$$

Case 2 gives the same result by raising to the odd power $(p+1)/2$. □

We can see by the work required in case (ii) of this proposition that the Gauss criterion is still not subtle enough to decide easily whether an odd prime q is a quadratic residue modulo p. To approach this final problem in quadratic residues the criterion is further refined to give Gauss' 'gem of higher arithmetic':

Theorem A.18 (Gauss' Quadratic Reciprocity Law). *If p, q are distinct odd primes, then*

$$(p/q)\,(q/p) = (-1)^{(p-1)(q-1)/4}.$$

Proof (tough but worth it). By the Gauss criterion,

$$(q/p) = (-1)^{\nu}$$

where ν is the number of integers a in $1 \leqslant a \leqslant (p-1)/2$ such that there exists an integer b satisfying

$$aq = bp + r, \qquad -p/2 < r < 0.$$

There can be at most one such integer b for each a, so we can rephrase the requirement: ν is the number of ordered pairs (a, b) of integers satisfying

$$1 \leqslant a \leqslant (p-1)/2, \tag{9}$$

$$-p/2 < aq - bp < 0. \qquad (10)$$

Now from (9) and (10) we can deduce

$$bp < aq + p/2 \leqslant (p-1)q/2 + p/2 < pq/2 + p/2 = p(q+1)/2.$$

Hence $b < (q+1)/2$, and (10) implies $b \geqslant 1$, so we have

$$1 \leqslant b \leqslant (q-1)/2. \qquad (11)$$

Since (9) and (10) imply (11), it does no harm to add (11) to the list of requirements to be satisfied by the ordered pair (a, b), so that ν is the number of pairs (a, b) of integers satisfying (9)–(11). It actually does a lot of good because of the symmetry of (9) and (11). Interchanging p, q, and a, b, we also have

$$(p/q) = (-1)^{\mu}$$

where μ is the number of ordered pairs of integers (a, b) satisfying (11), (9) and

$$-q/2 < bp - aq < 0 \qquad (10)'$$

which can be written

$$0 < aq - bp < q/2. \qquad (12)$$

Since p, q are distinct primes, (9) and (11) imply $aq - bp \neq 0$, so $\nu + \mu$ is the number of ordered pairs of integers (a, b) satisfying (9), (11) and

$$-p/2 < aq - bp < q/2. \qquad (13)$$

Now

$$(p/q)(q/p) = (-1)^{\nu+\mu}, \qquad (14)$$

so we only really require to know $\nu + \mu \mod 2$.
Let

$$R = \{(a, b) \in \mathbf{Z}^2 \mid 1 \leqslant a \leqslant (p-1)/2,\ 1 \leqslant b \leqslant (q-1)/2\}.$$

Then R has $(p-1)(q-1)/4$ elements. Now partition R as follows:

$$R_1 = \{(a, b) \in R \mid aq - bp \leqslant -p/2\}$$
$$R_2 = \{(a, b) \in R \mid -p/2 < aq - bp < q/2\}$$
$$R_3 = \{(a, b) \in R \mid q/2 \leqslant aq - bp\}.$$

Then R_2 is the set of solutions of (9), (11), (13) as required. The map $f: \mathbf{Z}^2 \to \mathbf{Z}^2$ given by $f(a, b) = ((p+1)/2 - a, (q+1)/2 - b)$ is easily seen to restrict to a bijection from R_1 to R_3 (check it!), hence $|R_1| = |R_3|$. This implies

$$|R| = |R_1| + |R_2| + |R_3|$$
$$\equiv |R_2| \qquad (\text{mod } 2)$$

so

$$(p-1)(q-1)/4 \equiv \nu + \mu \qquad (\text{mod } 2).$$

From (14) we see

$$(p/q)(q/p) = (-1)^{(p-1)(q-1)/4}.$$ □

As an immediate deduction we have the reciprocity law in the form given by Gauss:

Theorem A.19. *If p and q are distinct odd primes, at least one of which is congruent to* 1 *modulo* 4, *then p is a quadratic residue of p if and only if q is a quadratic residue of p; otherwise if neither is congruent to* 1 *modulo p then precisely one is a quadratic residue of the other.*

Proof. If at least one of p, q is congruent to 1 modulo 4, then $(p-1)(q-1)/4$ is even, so

$$(p/q)(q/p) = 1,$$

whence $(p/q) = +1$ if and only if $(q/p) = +1$. If neither is congruent to 1 modulo 4, then $(p-1)(q-1)/4$ is odd,

$$(p/q)(q/p) = -1,$$

so precisely one of (p/q), (q/p) is $+1$ and the other is -1. □

We can imagine Gauss' intense pleasure at discovering this remarkable result. To see its power, we only have to compare the easy way this resolves problems involving quadratic residues compared with the two criteria given earlier or with ad hoc methods. Its use is even more clear when allied with Legendre's clever symbol.

Example. Is 1984 a quadratic residue modulo 97?

$$(1984/97) = (44/97) = (2/97)^2(11/97) = (\pm 1)^2(11/97)$$
$$= (11/97) = (97/11) = (9/11) = (3/11)^2 = 1$$

(because $97 \equiv 1 \bmod 4$). Hence 1984 is a quadratic residue modulo 97.

By putting together the appropriate results of this chapter the question of whether a specific number k is a quadratic residue modulo m may be completely solved by a succession of reductions of the problem:

(i) By factorizing m and using proposition 10 which says k is a quadratic residue modulo m if and only if it is a quadratic residue modulo each prime factor of m.

(ii) If 2 is a prime factor, that part of the problem may be solved by Proposition A.11: if 2^e is the largest power of 2 dividing m, for $e = 1$, any odd k is a quadratic residue, for $e = 2$ we must check that $k \equiv 1 \pmod 4$ and for $e \geqslant 3$ we must check that $k \equiv 1 \pmod 8$.

(iii) For an odd prime factor p of m, we calculate (k/p). First we can reduce k modulo p so that we may assume $1 \leqslant k < p$. Then we factorize $k = q_1^{f_1} \ldots q_s^{f_s}$ where the q_i are primes and write

$$(k/p) = (q_1/p)^{f_1} \ldots (q_s/p)^{f_s}.$$

We only need consider (q_i/p) for f_i odd and since $q_i < p$, we can use Gauss reciprocity to obtain

$$(q_i/p) = (p/q_i)^2(q_i/p) = (p/q_i)(-1)^{(p-1)(q_i-1)/4}$$

This reduces the problem to calculating (p/q_i) where $p > q_i$, and reducing p modulo q_i, we then have to calculate Legendre symbols for smaller primes. Successive reductions of this nature lead to a complete solution.

Example. Is 65 a quadratic residue modulo 124?
Since $124 = 2^2 \times 31$, we must check if 65 is a quadratic resi-

due modulo 2^2 and 31. Modulo 2^2 we have $65 \equiv 1 \pmod 4$, so the answer is yes, by Proposition A.17(i). Modulo 31 we have

$$\begin{aligned}(65/31) &= (3/31) = (31/3)(-1)^{30 \times 2/4} = -(31/3) \\ &= -(1/3),\end{aligned}$$

and 1 is a quadratic residue modulo 3, so $(65/31) = -1$. Thus 65 is not a quadratic residue modulo 124.

It is the reduction of the seemingly complicated problem of quadratic residues to simple arithmetic such as this which highlights the brilliance of the jewel in Gauss's number-theoretic crown.

Exercises

A.1. Solve the following congruences (where possible)
 (i) $3x \equiv 14 \pmod{17}$,
 (ii) $6x \equiv 3 \pmod{35}$,
 (iii) $3x \equiv 13 \pmod{18}$,
 (iv) $20x \equiv 60 \pmod{80}$.

A.2. Solve the quadratic congruences (where possible):
 (i) $3x^2 + 6x + 5 \equiv 0 \pmod 7$,
 (ii) $x^2 + 5x + 3 \equiv 0 \pmod 4$,
 (iii) $x^2 \equiv 1 \pmod{12}$,
 (iv) $x^2 \equiv 0 \pmod{12}$,
 (v) $x^2 \equiv 2 \pmod{12}$.

A.3. Let $a_1, a_2, \ldots, a_n$ be the complete set of residues modulo n (not in any specific order), and let b be an integer relatively prime to n and c any integer. Show that

$$a_1b + c, a_2b + c, \ldots, a_nb + c$$

is also a complete set of residues.

A.4. Calculate the Euler function $\phi(n) = |U(\mathbf{Z}_n)|$ for $n = 4, 6, 12, 18$.

A.5. Determine all the generators of $U(\mathbf{Z}_7)$, $U(\mathbf{Z}_{11})$, $U(\mathbf{Z}_{17})$.

A.6. Show that there exist primitive roots s modulo p such that

$$s^{p-1} \not\equiv 1 \pmod{p^2}.$$

A.7. Calculate the exponents of the following groups: $U(\mathbf{Z}_4)$, $U(\mathbf{Z}_6)$, $U(\mathbf{Z}_8)$, $U(\mathbf{Z}_{10})$. Which of these groups are cyclic?

A.8. Show that for an odd prime p there are as many square residue classes as non-squares in $U(\mathbf{Z}_{p^e})$ for $e \geqslant 1$.

A.9. Show that in $U(\mathbf{Z}_{2^e})$ there are exactly 2^{e-3} squares and 3.2^{e-3} non-squares for $e \geqslant 3$.

A.10. Determine the squares in the following groups: $U(\mathbf{Z}_7)$, $U(\mathbf{Z}_{12})$, $U(\mathbf{Z}_{49})$.

A.11. Use Euler's criterion to check whether 7 is a quadratic residue modulo 23. Answer the same question using Gauss' criterion. Now calculate $(7/23)$ using Gauss reciprocity.

A.12. Compute the following Legendre symbols: $(-1/179)$, $(6/11)$, $(2/97)$.

A.13. Compute $(97/1117)$, $(2437/811)$, $(23/97)$.

A.14. Is 1984 a quadratic residue modulo 365?

A.15. Is 2001 a quadratic residue modulo 1820?

A.16. Find the primes for which 11 is a quadratic residue.

A.17. Define the Jacobi symbol (k/m) for relatively prime integers k, m $(m > 0)$ by factorizing $m = p_1^{e_1} \ldots p_r^{e_r}$ and writing

$$(k/m) = (k/p_1)^{e_1} \ldots (k/p_r)^{e_r}.$$

If k, r are prime to m and $k \equiv r \pmod{m}$, show

$$(k/m) = (r/m).$$

A.18. If (k/m) is the Jacobi symbol of question 17, for m positive and odd, prove

$$\text{(i)} \quad (-1/m) = (-1)^{(m-1)/2},$$

$$\text{(ii)} \quad (2/m) = (-1)^{(m^2-1)/8}.$$

For k, m positive and relatively prime, prove

$$(k/m)(m/k) = (-1)^{(k-1)(m-1)/4}.$$

References

[GT] I. N. Stewart (1973). *Galois Theory*. Chapman and Hall, London.

[HH] B. Hartley and T. O. Hawkes (1970). *Rings, Modules, and Linear Algebra.* Chapman and Hall, London.

Further reading on algebraic number theory

1. Z. I. Borevič and I. R. Šafarevič (1966). *Number Theory*. Academic Press, New York.
2. J. W. S. Cassels and A. Fröhlich (1967). *Algebraic Number Theory*. Academic Press, New York.
3. H. Cohn (1978). *A Classical Invitation to Algebraic Numbers and Class Fields,* Springer, Berlin and New York.
4. H. M. Edwards (1977). *Fermat's Last Theorem,* Springer, Berlin and New York.
5. G. H. Hardy and E. M. Wright (1954). *An Introduction to the Theory of Numbers,* Oxford University Press, Oxford.
6. S. Lang (1970). *Algebraic Number Theory.* Addison-Wesley, Reading Massachusetts.
7. O. T. O'Meara (1973). *Introduction to Quadratic Forms.* Springer, Berlin and New York.
8. H. Pollard (1950). *The Theory of Algebraic Numbers.* Mathematical Association of America, Buffalo.
9. P. Ribenboim (1972). *Algebraic Numbers.* Wiley-Interscience, New York.
10. P. Ribenboim (1979). *13 Lectures on Fermat's Last Theorem,* Springer, Berlin and New York.

11. P. Samuel (1967). *Théorie Algébrique des Nombres.* Hermann, Paris; English translation Houghton Mifflin, New York (1975).
12. A. D. Thomas (1977). *Zeta-functions: an Introduction to Algebraic Geometry.* Research Notes in Mathematics **12**, Pitman, London and San Francisco.
13. A. Weil (1967). *Basic Number Theory.* Springer, Berlin and New York.
14. H. Weyl (1940). *Algebraic Theory of Numbers.* Princeton University Press, Princeton NJ.

Other material

15. T. M. Apostol (1957). *Mathematical Analysis.* Addison-Wesley, Reading Massachusetts.
16. A. Baker (1966). Linear forms in the logarithms of algebraic numbers. *Mathematika* **13**, 204–216.
17. B. J. Birch (1968). Diophantine analysis and modular functions. *Proceedings of the Conference on Algebraic Geometry,* Tata Institute Bombay 35–42.
18. S. Bloch (1984). The Proof of the Mordell Conjecture, *Mathematical Intelligencer* **6**, no. 2, 41–47.
19. J. Brillhart, J. Tonascia and P. Weinberger (1971). On the Fermat Quotient, in *Computers in Number Theory.* Academic Press. New York, 213–222.
20. A. Cauchy (1897). *Oeuvres* 1(X) 276–285, 296–308. Gauthier-Villars, Paris.
21. H. Chatland and H. Davenport (1950). Euclid's algorithm in real quadratic fields. *Canadian Journal of Mathematics* **2**, 289–296.
22. M. Deuring (1968). Imaginäre quadratischen Zahlkörper mit der Klassenzahl Eins. *Inventiones Mathematicae* **5**, 169–179.
23. H. M. Edwards (1975). The background of Kummer's proof of Fermat's Last Theorem for regular primes. *Archive for the History of Exact Sciences* **14**, 219–236.
24. C. F. Gauss (1966). *Disquisitiones Arithmeticae.* (Translated by A. A. Clarke), Yale University Press.
25. D. M. Goldfeld (1985). Gauss' class number problem for imaginary quadratic fields, *Bulletin of the American Mathematical Society* **13**, 23–37.
26. E. S. Golod and I. R. Šafarevič (1964). On class field towers, *Izvestija Akademii Nauk SSSR Serija Matematičeskaja* **28**, 261–272; *American Mathematical Society Translations, 2nd Series,* **48** (1965) 91–102.

27. G. H. Hardy (1960). *A Course of Pure Mathematics.* Cambridge University Press.
28. K. Heegner (1952). Diophantische Analysis und Modulfunktionen. *Mathematische Zeitschrift* **56**, 227–253.
29. H. Heilbronn and E. H. Linfoot (1934). On the imaginary quadratic corpora of class-number one. *Quarterly Journal of Mathematics (Oxford)* **5**, 293–301.
30. K. Inkeri (1947). Über den Euklidischen Algorithmus in quadratischen Zahlkörpern, *Annales Academiae Scientiarum Fennicae* **41**, pp. 35.
31. W. Ledermann (1973). *Introduction to Group Theory.* Oliver and Boyd, Edinburgh and London.
32. I. D. Macdonald (1968). *The Theory of Groups,* Oxford University Press.
33. W. Narkiewicz (1967). Class number and factorization in quadratic number fields. *Colloquium Mathematicum* **17**, 167–190.
34. C. Reid (1970). *Hilbert,* Springer, Berlin and New York.
35. P. Samuel (1971). About Euclidean rings, *Journal of Algebra* **19**, 282–301.
36. J. L. Selfridge and B. W. Pollock (1967). Fermat's Last Theorem is true for any exponent up to 25,000. *Notices of the American Mathematical Society* **11**, 97, abstract no. 608–138.
37. C. L. Siegel (1968). Zum Beweise des Starkschen Satzes. *Inventiones Mathematicae* **5**, 169–179.
38. H. M. Stark (1967). A complete determination of the complex quadratic fields of class-number one. *Michigan Mathematical Journal* **14**, 1–27.
39. I. N. Stewart and D. O. Tall (1977). *The Foundations of Mathematics,* Oxford University Press, Oxford.
40. D. J. Struik (1962). *A Concise History of Mathematics,* Bell, London.
41. E. C. Titchmarsh (1960). *The Theory of Functions,* Oxford University Press, Oxford.
42. S. Wagstaff (1978). The irregular primes to 125000, *Mathematics of Computation* **32**, 583–591.

Index